% Error = |STV - EXP| / STV × 100

EXPERIMENTS IN PHYSICAL SCIENCE

Second Edition

George N. Caviris **Dr. Sophie Moore**

State University of New York
Farmingdale

Illustrations by
Zevelyn Caviris

KENDALL/HUNT PUBLISHING COMPANY
4050 Westmark Drive Dubuque, Iowa 52002

ISBN 978-0-7872-1389-3

Printed in the United States of America
10

Preface

The second edition has numerous changes and incorporates several new features into the laboratory manual. These include the following:

(a) Numerous worked out examples have been introduced to many experiments - especially to those dealing with topics not usually covered in the theory of a physical science course (e.g., significant figures, graphs).

(b) The number of figures has been greatly increased to help the student visualize the discussions in this manual.

(c) Several new experiments have been added dealing with introductory topics; Experiment 1 introduces the student to scientific notation, the metric system and simple measurements; Experiment 2 teaches significant figures and obtaining the units for calculated quantities; Experiment 4 explains linear graphs; and Experiment 6 uses the simple pendulum to illustrate how to analyze a non-linear function.

(d) Experiment 4, *Linear Graphs,* does not involve gathering data. If performed during the first laboratory class it is possible to assign it to a student who registered late without having to schedule a make-up laboratory. If the instructor wishes to have his or her class obtain data, he or she may have the students do that portion of Experiment 5, *The Conversion of Inches to Centimeters,* that involves analyzing a linear graph.

(e) There are new experiments in kinematics (Experiments 13 and 14), statistics (Experiment 20), vector addition (Experiment 24), determining absolute zero (Experiment 29), and mineral identification (Experiment 41).

(f) Questions have been added to the ends of most experiments.

We wish to thank our colleagues in the Physics Department at SUNY Farmingdale for their suggestions, criticisms, and support: Dr. Loyd Makarowitz (Chairman), Raymond Bigliani, Dr. Yuen Chinn, Dr. Paul Kramer, Bernard Malara, Edmund Mayer, Gustave Nelson, Dr. Peter Nolan, Morton Seitelman. We are especially grateful to Soloman Ayo, the Senior Laboratory Technician for the department.

The authors wish to thank Zevelyn Caviris for the many illustrations that appear throughout this manual.

George Caviris
Sophie Moore

Contents

EXPERIMENT 1

SCIENTIFIC NOTATION, THE METRIC SYSTEM, SIMPLE MEASURING DEVICES AND DERIVED QUANTITIES

Objectives

1. To learn scientific notation
2. To learn the metric system
3. To learn how to measure length with a meter stick (ruler)
4. To learn how to measure mass with a balance.
5. To learn about derived quantities
6. To learn how to calculate the weight of a mass

Equipment

A meterstick and a balance.

Discussion and Theory

Scientific Notation

Quite often the numbers used in scientific work are either very large or very small and require writing many zeros unless some compact method of writing these numbers is used. *Scientific Notation* is such a method. To understand this notation, one needs to be familiar with the powers of ten. We may write:

10 as	10^1
100 as 10x10 or	10^2
1,000 as 10x10x10 or	10^3
.	
.	
.	
1,000,000 as 10x10x10x10x10x10 or	10^6
etc.	

Note that for the powers of ten, the exponent of the ten is the same as the number of zeros when the number is written out in the conventional method. The number one has no zeros after it and its exponent when it is written as powers of ten is zero (i.e., $1 = 10^0$).

For numbers less than one, we may write:

0.1 as 1/10 or	10^{-1}
0.01 as 1/(10x10) or	10^{-2}
0.001 as 1/(10x10x10) or	10^{-3}
.	
.	
.	
0.0000001 as 1/(10x10x10x10x10x10x10) or	10^{-7}
etc.	

Here the exponent is a negative number and is always one more than the number of zeros before the one when it is written in conventional form.

In scientific notation numbers are written as a number between one and 9.99999... multiplied by ten raised to some power. Therefore, in scientific notation a number is written in the form

$$\text{x.xx...} \times 10^n \qquad (1\text{-}1)$$

Examples

$$3{,}102{,}000{,}000{,}000 = 3.102 \times 10^{12}$$
$$712.345 = 7.12345 \times 10^2$$
$$60{,}000{,}000 = 6.00 \times 10^7$$
$$0.00000004100 = 4.100 \times 10^{-8}$$
$$0.0031128 = 3.1128 \times 10^{-3}$$

Scientific notation permits one to express very large and very small numbers in a more compact and convenient form. As an example of where scientific notation may be necessary, consider the speed of light written as 299,800,000 m/s. Some calculators do not display 9 digits and cannot accommodate this number in this form. However, a scientific calculator is able to handle this number if it is expressed in scientific notation as 2.998×10^8 m/s. Similarly, the mass of an electron, 9.11×10^{-31} kg, can only be handled by a scientific calculator in scientific notation.

The Metric System

The international system of units, or the SI system as it is called (from the French name of the sysem: "Systeme International"), uses length, mass and time as its fundamental mechanical quantities from which all other mechanical quantities are defined. The *meter* (m) is used to measure length, the *kilogram* (kg) is used to measure mass and the *second* (s) is used to measure time. These units are used to define all other mechanical metric units.

In the metric system of units, different units for each quantity differ by multiples of 10 or 0.1 of the base unit The exact multiple is obtained from a prefix which is added to the name of the reference unit. Table 1-1 lists some of the more common metric prefixes. The student should learn these prefixes and how they are used.

Table 1-1
Common Metric Prefixes

Prefix	Abbreviation	Multiple as a Decimal	Multiple as a Power of Ten
micro	μ	0.000001	10^{-6}
milli	m	0.001	10^{-3}
centi	c	0.01	10^{-2}
deci	d	0.1	10^{-1}
deka	da	10.	10^{1}
kilo	k	1,000.	10^{3}
mega	M	1,000,000.	10^{6}

The use of these prefixes is best seen by considering a number of examples.

Examples

1 mm = 0.001 m = 10^{-3} m
1 cm = 0.01 m = 10^{-2} m
1 km = 1000 m = 10^{3} m
1 mg = 0.001 g = 10^{-3} g
1 cg= 0.01 g = 10^{-2} g
1 kg = 1000 g = 10^{3} g
1 ms = 0.001 s = 10^{-3} s
1 cs= 0.01 s = 10^{-2} s
1 ks = 1000 s = 10^{3} s

8.21 mm = 8.21 x 10^{-3} m
15 cm = 15 x 10^{-2} m = 0.15 m
64.1 km = 64.1 x 10^{3} m = 6.41 x 10^{4} m
15 mg = 15 x 10^{-3} g = 1.5 x 10^{-2} g
0.89 cg = 0.89 x 10^{-2} g = 8.9 x 10^{-3} g
18.3 kg = 18.3 x 10^{3} g = 1.83 x 10^{4} g
0.1 ms = 0.1 x 10^{-3} s = 1 x 10^{-4} s
3.45 cs = 3.45 x 10^{-2} s
253 ks = 253 x 10^{3} s = 2.53 x 10^{5} s

The Ruler and the Meter Stick

The simplest instrument for measuring length is a straight edge with a scale drawn along it. Such an instrument is called a *ruler.*

A *meterstick* is a one meter long stick that is usually subdivided along one or both lengths of the stick into centimeters and millimeters.

Whenever a ruler is used, the student should know into which units the ruler is subdivided and make certain that the zero on the ruler is properly positioned before the length of any object is read from the ruler. As with any measurement, *make certain to include the units when the measurement is recorded.*

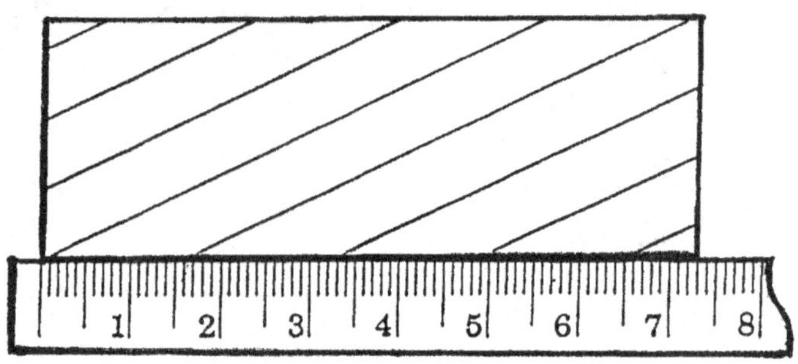

Figure 1-1 Measuring Length with a Ruler

Example

To determine the length of one side of the object shown in Figure 1-1 in metric units use a ruler that shows centimeters and millimeters (as subdivisions of the centimeters). *Position the ruler along the edge to be measured so that the zero falls exactly where the edge of the object starts.* The length of the object is then read at that point along the ruler where the object ends. In this case, the length is 7.3 cm or 73 mm.

Measuring Mass With a Balance

One of the most commonly used instruments in the laboratory is the *balance*. It is used to measure mass. One type of balance is shown in Figure 1-2. The object whose mass, M, is to be determined is placed on platform A on the zero side of the rider scale. The mass is determined by adding masses to platform B and moving the rider until the balance indicator points to zero (the center position). If m_r is the reading on the rider scale and m_b is the mass on platform B when the balance indicator reads zero, then the mass, M, of the object on platform A is

$$M = m_b + m_r. \tag{1-2}$$

Example

In Figure 1-2 the balance indicator is at zero. The mass on platform B is

$$m_b = 200\ g + 50\ g = 250\ g.$$

The reading m_r on the rider is 6.1 g. The mass on platform A is

$$M = m_b + m_r = 250\ g + 6.1\ g = 256.1\ g.$$

There are several variations of the balance. A plane platform balance will not have any beam with the rider scale. A multi-beam balance will have several riders that need to be adjusted to obtain balance.

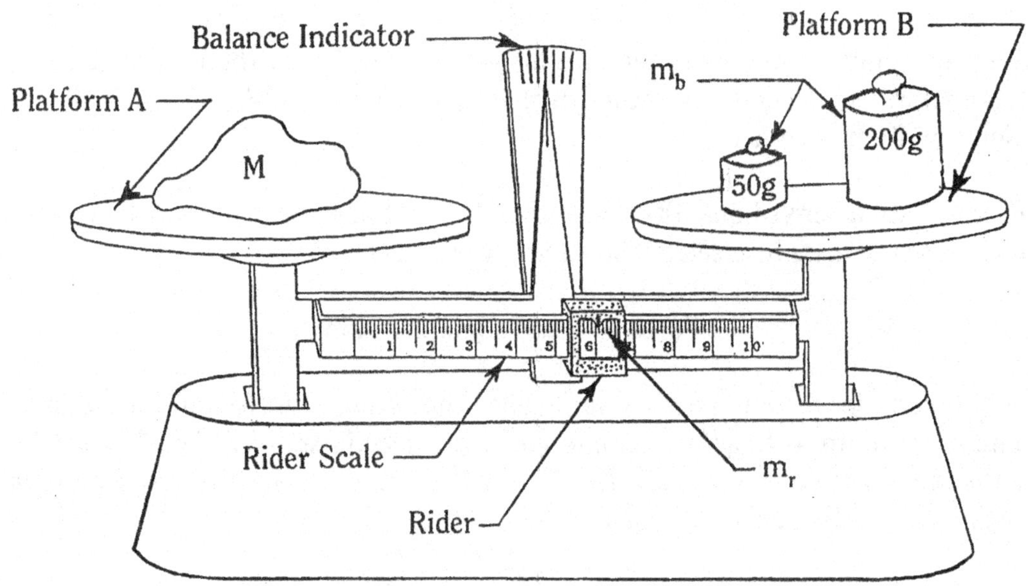

Figure 1-2 Measuring Mass with a Balance

Derived Quantities

Quantities defined from other quantities are called *derived quantities*. The units used to measure derived quantities are called *derived units*. By definition,

$$\text{Speed} = \frac{\text{Distance}}{\text{Time}} \qquad (1\text{-}3)$$

The speed is a an example of a derived quantity, since its definition is based on other quantities (the distance and the time). The units used to measure speed, such as m/s and cm/s, are the derived units.

Derived units that are expressed in terms of the fundamental SI units (meters, kilograms and seconds) are also called SI units. The unit "m/s" is the SI unit for speed. Units that may be expressed in terms of *c*entimeters, *g*rams and *s*econds are called *cgs units*. The unit "cm/s" is the cgs unit for speed.

The acceleration is another example of a derived quantity. By definition,

$$\text{Acceleration} = \frac{\text{Change in Velocity}}{\text{Elapsed Time}} \qquad (1\text{-}4)$$

The unit for acceleration consists of the unit for velocity divided by the unit for time. In SI units the unit is meters per second per second which is written as m/s^2. In cgs units, the unit for acceleration is written as cm/s^2.

Force is also a derived quantity. According to Newton's Second Law, when a force acts on a mass it will cause the mass to accelerate in such a way that the

$$\text{Force} = (\text{mass}) \times (\text{acceleration}). \qquad (1\text{-}5)$$

In SI units, the unit for force is a derived unit called the *Newton*. A Newton is that force which will cause a one kilogram mass to accelerate one meter per second squared (i.e., $1\ N = 1\ kg\ m/s^2$). In cgs units the unit for force is the *dyne*. The dyne is that force which will cause a one gram mass to accelerate one centimeter per second squared (i.e., $1\ dyne = 1\ g\ cm/s^2$).

Weight

The *weight* of an object is the force of gravity acting on it. When an object is dropped its acceleration will be the acceleration due to gravity. From equation (1-5) we see that the weight, W, of an object can be calculated from

$$W = mg \qquad (1\text{-}6)$$

where m is the mass and g is the acceleration due to gravity. On the surface of the earth the acceleration due to gravity is

$$\begin{aligned} g &= 9.8 \text{ m/s}^2 \\ &= 980 \text{ cm/s}^2 \\ &= 32 \text{ ft/s}^2. \end{aligned}$$

Examples

(a) The weight of a 10 kilogram mass is $W = mg = (15 \text{ kg})(9.8 \text{ m/s}^2) = 147$ N.
(b) The weight of a 3.5 gram mass is $W = mg = (3.5 \text{ g})(980 \text{ cm/s}^2) = 3{,}430$ dyne.

Even though the weight of an object may be calculated from its mass, it is important for the student to realize that *the mass and the weight of an object are two different physical quantities.* The mass of an object remains the same everywhere in the universe. The weight of an object varies depending on its location in the universe.

Example

The weight of a 10 kg mass on the earth is $(10 \text{ kg})(9.8 \text{ m/s}^2) = 98$ N. On the moon, where the acceleration due to gravity is 1.6 m/s^2, the weight is $(10 \text{ kg})(1.6 \text{ m/s}^2) = 16$ N. The mass remains 10 kg at these locations.

Procedure

1. With a meterstick measure the length of ten objects. Your instructor may select specific objects to be measured. If you are given a free choice of objects to measure, choose objects of various lengths. Record your measured values in Table 1-2. Convert your measured values to all the units indicated. Express values greater then 100 and smaller then 0.1 in scientific notation.

2. With a balance determine the mass of ten objects. Your instructor may select specific masses to be measured. If you are given a free choice of masses to measure, do not choose objects with similar masses. Record your measured values in Table 1-3. Convert your measured values to all the units indicated. Express values greater then 100 and smaller then 0.1 in scientific notation.

3. Calculate the weight in Newtons of each of the ten masses. Record your results in Table 1-3.

Questions

1. Examine a 12 inch ruler. How are the inches subdivided? What is the smallest division?

2. Express the following in scientific notation: (a) 132.45, (b) 0.0000654, (c) 9,461,000, (d) 8,230, (e) 12, (f) 0.000005, (g) 0.10003, (h) 100.2, (i) 0.129, (k) 960,000,000,000.

3. Convert each of the following quantities to the units indicated by filling in the blanks. Express your results in scientific notation when the number is greater then 10 or less then 1.0.

(a) 5.3 m = ____________________ mm	(g) 3,500 m = ____________________ Mm
(b) 500 g = ____________________ kg	(h) 16.1 kg = ____________________ g
(c) 23.5 cm = ____________________ m	(i) 23.8 cg = ____________________ mg
(d) 600.5 m = ____________________ cm	(j) 0.00345 Mg = ____________________ kg
(e) 452 ms = ____________________ s	(k) 1.5 km = ____________________ mm
(f) 864 dm = ____________________ m	(l) 0.987 dag = ____________________ g

4. Express 10 kiloseconds in (a) minutes, (b) hours, (c) days, (d) years.

5. What is the weight of each of the following masses in Newtons and dynes: (a) 2.0 kg, (b) 0.02 kg, (c) 24 g, (d) 10 g.

Name______________________ Date Performed______________________
Lab Partners______________________ Date Submitted______________________

Table 1-2

Measured Lengths Expressed in Various Metric Units

	Measured item	Length in millimeters	Length in centimeters	Length in meters	Length in kilometers
1.					
2.					
3.					
4.					
5.					
6.					
7.					
8.					
9.					
10.					

Table 1-3

Measured Masses Expressed in Various Metric Units and Their Weights

	Measured item	Mass in milligrams	Mass in grams	Mass in kilograms	Weight in Newtons
1.					
2.					
3.					
4.					
5.					
6.					
7.					
8.					
9.					
10.					

EXPERIMENT 2

UNDERSTANDING ERRORS AND UNITS

Objectives

1. To understand how errors affect the results in experiments
2. To learn about significant figures in calculations
3. To learn how to calculate the percent error
4. To learn how to determine the units of calculated results

Equipment

A meter stick or a metric tape measure.

Discussion and Theory

Errors and Calculated Results

Whenever a measurement is made, its accuracy is limited. A limit on the precision of a measurement is imposed by the limitations of the measuring instrument. All laboratory measurements are assumed to have a certain amount of error. This error affects all results that are calculated from that measurement.

Significant Figures

In measuring the length of an object with a meter stick, one obtains a reading of 0.284 meters. If this meter stick can be read only to within an accuracy of 0.001 meters, then the length of the object falls between 0.283 and 0.285 meters. This measurement is said to have three *significant figures*. The significant figures are the digits 2,8 and 4. The first two digits are assumed to be accurately known, while the last digit, the 4, is assumed to be estimated. An estimation of any of the digits beyond the third digit is not possible.

Difficulties often arise with the digit zero. When a number starts with zeros these zeros are called leading zeros (e.g., the number .0023 has two leading zeros). *Leading zeros are never significant.*

Examples

The number 0.0000586 has three significant figures and 0.0000001 has only one significant figure.

Any zeros between non-zero digits are called imbedded zeros (e.g., the number 90.1003 has three imbedded zeros and the number 0.00001006 has two imbedded zeros). *Imbedded zeros are always significant.*

Examples

The number 8.016 has four significant figures, 10,103.02 has seven significant figures and 0.0010211 has five significant figures.

When a number ends with zeros these zeros are called trailing zeros. *Trailing zeros are significant only if one or more of the trailing zeros is to the right of the decimal point.*

Examples

The number 12.300 has five significant figures and 900.0 has four significant figures.

Trailing zeros present a problem when it comes to significant figures when they fall to the left of the decimal point. Trailing zeros are often needed as place holders and may or may not be indicative of the precision of a measurement.

Example

A measurement of 320 meters has only two significant figures if the measurement is made to the closest ten meters, while it has three significant figures if the measurement is made to the closest meter.

A way around this problem is to use scientific notation. When scientific notation is used to express a result, all the digits shown before the "x 10^n" are significant. Care should be taken to indicate the appropriate accuracy of the result written. Here trailing zeros should be used to indicate the proper number of significant figures when needed.

Example

If a measurement of 3,200 meters is made to the nearest meter it should be written as 3.200 X 10^3 m to indicate that the result is accurate to four significant figures. If this measurement is made to the nearest ten meters it should be written as 3.20 X 10^3 m to indicate that the result is accurate to three significant figures. If this measurement is made to the nearest 100 meters it should be written as 3.2 X 10^3 m to indicate that the result is accurate to only two significant figures.

Determination of the Number of Significant Figures in Calculated Results

Whenever measurements are accurate to a limited number of significant figures, the number of significant figures in any result calculated from these measurements is also limited.

Example

Consider two measured quantities 1.2*3* cm and 15.8*5* cm. In each of these the last digit is written in underlined italics to indicate it is estimated. If we multiply these numbers together we get

```
  15.85
 X1.23
  4755
 3170
1585
19.4955
```

Answer: 19.5 cm^2

In this multiplication the effect of the uncertainty of the last digit of each of the two multiplied numbers has been shown in underlined italics. As can be seen, all the digits to the right of the decimal point in the result have an uncertainty associated with them. Therefore, the .4955 is rounded off to .5 and the final result is 19.5 cm. The final result has only three significant figures.

In this laboratory, the following rules will be used in determining the number of significant figures resulting from a series of calculations:

1. Whenever numbers are added or subtracted, the last significant figure of the answer is in the same column as the last significant figure of the least precise number(s). The answer is then rounded off to end in this column.

Examples

(a)

```
 123.163
  21.4
+11.123
 155.686
```

Answer: 155.7

(b)

```
876.548
  -1.12
875.428
```

Answer: 875.43

In example (a) the final result is accurate only to the nearest tenth since 21.4 is precise to the nearest tenth. In example (b) the final result is accurate only to the nearest hundredth since -1.12 is precise to the nearest hundredth.

2. When 2 or more numbers are multiplied or divided, the number of significant figures in the answer is the same as that of the number that has the least number of significant figures.

Examples

(a) $$\frac{9.62}{3.1848} = 3.02$$

(b) $$\frac{(3.2222 \times 10^{-5})}{(3.1 \times 10^{5}) \times (1.34)} = 7.8 \times 10^{-11}$$

In example (a) there are only three significant figures in the number with the least number of significant figures,9.62. Therefore, the result must have thee significant figures. In example (b) there are only two significant figures in the number with the least number of significant figures, 3.1×10^{-5}. Therefore, the result must have two significant figures.

Percent Error

When we calculate a result in the laboratory, we will find that it often varies from the standard or theoretical value. This variation is defined as the *percent error*. The percent error is

$$\% \text{ Error} = \frac{|\text{Standard Value - Experimental Value}|}{(\text{Standard Value})} \times 100 . \qquad (2\text{-}1)$$

The two vertical lines "| ... |" are used to indicate the absolute value of the difference between the standard and experimental values. This means that the difference is always taken as positive, even if the experimental value is larger then the standard value. Therefore, the percent error is always expressed as a positive number.

Example

An experimental value of 995 cm/s^2 is obtained for the acceleration due to gravity. The standard value is 980 cm/s^2. The percent error is

$$\% \text{ error} = \frac{|980 - 995|}{980} \times 100 = 1.53\% .$$

Even though 995 cm/s^2 is larger than 980 cm/s^2 the percent error is written as a positive number because the absolute value is defined to be always positive.

Units In Calculated Results

All physical quantities have units. If you measure the length of an object to be 18 units long, your measurement would not mean very much to someone else if he or she does not know with what units you made the measurement. There is a big difference between a length of 18 centimeters and 18 meters or 18 kilometers.

Whenever a student makes a measurement, he or she must record it with the proper units.

When calculations are made using physical quantities the calculated results also have units. *The units are a part of any calculated quantity and must be included in the results.*

When calculations are made with physical quantities, the units are added, subtracted, multiplied, divided etc. in the same manner as the physical quantities.

Examples

(1) A rectangle is 5.20 cm wide and 3.31 cm long. What is the area?

Area = (length) X (width) = (5.20cm) X (3.31cm) = (5.20 X 3.31)(cm X cm) = 17.2 cm^2.

(2) What is the volume of a crate 3.10 m long, 2.40 m wide and 1.80 m high.

Volume = (length) X (width) X (height) = (3.10 m) X (2.40 m) X (1.80 m)

= (3.10 X 2.40 X 1.80) (m X m X m) = 13.4 m^3.

(3) A sphere has a radius of 0.932 cm. What is its volume?

Volume = $(4/3)\pi r^3 = (4/3)\pi(0.932cm)^3 = (4/3)\pi(0.932)^3(cm)^3 = 3.39\ cm^3$.

(*Note*: $(4/3)\pi$ is not a measured quantity. Therefore, it imposes no restriction on the number of significant figures)

(4) The above sphere has a mass of 28.15 grams. What is its density?

Density = Mass/Volume = 28.15g/3.39cm^3 = 8.30 g/cm^3.

Procedure

1. Measure the length and width of an 8 1/2" by 11" piece of paper in centimeters. Measure the length and width in metric units of nine other rectangular objects in the laboratory (e.g., a table top, a door, the floor of the laboratory, etc.). Record your measurements for each to the proper number of significant figures in Table 2-1. Do not forget to indicate all units.

2. Calculate the areas of the objects measured by multiplying the length by the width of each. Record your results to the proper number of significant figures in Table 2-1. Do not forget to include all units. Indicate the number of significant figures in your results in the space provided.

3. Use 603.22 cm^2 as the standard value for the area of the 8 1/2" by 11" piece of paper. Calculate and record in Table 2-1 the percent error of your experimental result for this area.

Questions

1. Did all the measurements that you made in this experiment have the same number of significant figures?

2. Calculate the following and give your answer to the proper number of significant figures. Include units in all answers. Express your answer in scientific notation if the answer is greater then 10 or less then 1. (a) (1.3 cm) x (1.00 cm), (b) 2.000 ft x 12.3 ft, (c) 1.1 kg x (9.00 m / s^2), (d) 220 g / (11 cm^3), (e) $(99.\ kg)^3$, (f) $(2.00 \times 10^3\ cm)^2$, (g) $(2.00 \times 10^3 kg)(1.0 \times 10^2 kg) / (5.00m)^2$, (h) $(8.0 \times 10^8\ m) \times (20 \times 10^{11}\ m)$.

3. What is the percent error for an experimental result of 968 m/s^2 if the standard value is 980 m/s^2

4. Calculate the percent error for each of the following:

Standard Value	Experimental Value
2.54 cm	2.59 cm
330 m/s	324 m/s
9.8 m/s^2	10.2 m/s^2

Name____________________ Date Performed____________________
Lab Partners ____________________ Date Submitted____________________
____
____

Table 2-1

Determination of Areas to the Proper Number of Significant Figures

	Object measured	Length	Width	Area	Number of Significant Figures
1.	8 1/2" by 11" paper	______	______	______	______
2.	______	______	______	______	______
3.	______	______	______	______	______
4.	______	______	______	______	______
5.	______	______	______	______	______
6.	______	______	______	______	______
7.	______	______	______	______	______
8.	______	______	______	______	______
9.	______	______	______	______	______
10.	______	______	______	______	______

Percent error of experimental area of 8 1/2" by 11" paper = ____________________

EXPERIMENT 3

THE SCIENTIFIC CALCULATOR

Objective

To learn to use a scientific calculator.

Equipment

A scientific calculator.

Procedure

The use of your scientific calculator is discussed in detail in its instruction manual. The following is a list of problems to familiarize you with the calculator. Results greater then 100 and smaller then 0.1 should be expressed in proper scientific notation. Values for the trigonometric functions should be expressed to 4 significant figures. Angles obtained from the inverse trigonometric functions should be rounded to the nearest tenth of a degree. If your instructor has discussed significant figures, express your answers to the proper number of significant figures. If your instructor covers significant figures at some future date, he or she may then have you rewrite your answers to the proper number of significant figures in the extra column provided.

EXPERIMENT 4

LINEAR GRAPHS

Objectives

1. To learn about straight line graphs
2. To learn how to draw straight line graphs
3. To learn how to analyze straight line graphs

Discussion and Theory

The relationship between two physical quantities is often determined by the analysis of a graph. A well laid out graph will clearly show the variation of one physical quantity as a function of another.

Usually graphs are plotted on a coordinate system which consists of two perpendicular axes. The horizontal axis is referred to as the *x-axis* and the vertical axis as the *y-axis*. The quantity on the x-axis is called the *independent variable*. The quantity on the y-axis is called the *dependent variable*. A graph shows the variation of the dependent variable as the independent variable changes.

Straight Line Graphs - Linear Functions

The simplest kind of graph is the straight line. The equation for a straight line is

$$y = mx + b, \tag{4-1}$$

where m is called the slope and b is called the *y-intercept*. A mathematical function that has the form of equation (4-1) is said to be linear.

Examples

(1) The function $y = 2x + 5$ is linear. It has a slope of 2 and a y-intercept of 5.

(2) The function $y = -x + 4$ is linear. It has a slope of -1 and a y-intercept of 4.

(3) The function $R = 3.4S - 10$ has a slope of 3.4 and a y-intercept of -10. Notice that the variables do not need to be represented by x and y for the function to be linear. What is important is that the equation has the form of equation (4 -1).

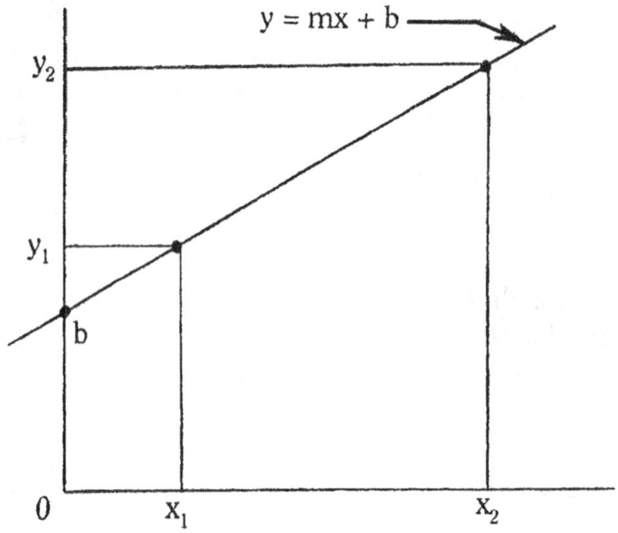

Figure 4-1 The Graph of a Straight Line

Figure 4-1 is the graph of a straight line. As we increase the quantity on the x-axis from x_1 to x_2 we see that the quantity on the y-axis changes from y_1 to y_2. From equation (4-1), $y_1 = mx_1 + b$ and $y_2 = mx_2 + b$. The total change in y is

$$\begin{aligned} y_2 - y_1 &= (mx_2 + b) - (mx_1 + b) \\ &= mx_2 - mx_1 + b - b \\ &= m(x_2 - x_1). \end{aligned}$$

If we solve for the slope we obtain

$$m = \frac{(y_2 - y_1)}{(x_2 - x_1)} . \tag{4-2}$$

Usually the change in y is denoted by the quantity $\Delta y = (y_2 - y_1)$ and the change in x by the quantity $\Delta x = (x_2 - x_1)$. Substitution into equation (4-2) results in the form

$$m = \frac{\Delta y}{\Delta x} = \text{the slope.} \tag{4-3}$$

The y-intercept, b, is the point where the line crosses the y-axis. This occurs at $x = 0$.

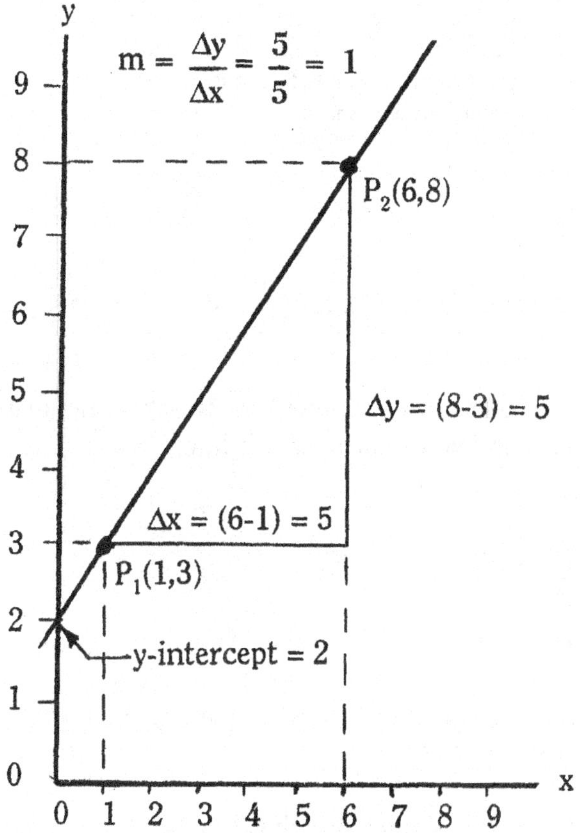

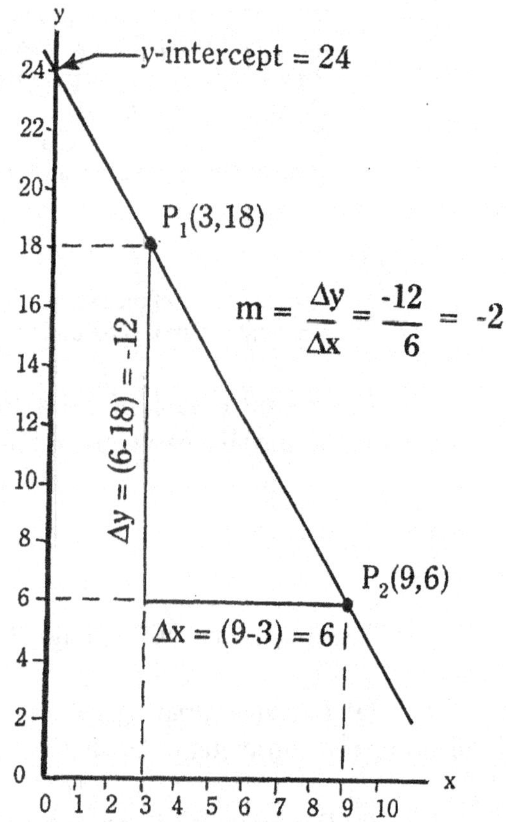

Figure 4-2 Straight Line Graphs with Positive and Negative Slopes

Examples

(1) In Figure 4-2(a) the coordinates of two points P_1 and P_2 are determined as shown. We see that $\Delta x = 5$ and $\Delta y = 5$. The slope is obtained from $m = \Delta y/\Delta x = 5/5 = 1$. The y-intercept is 2. The equation of this line is obtained by substituting these values into equation (4-1). This gives

$$y = x + 2.$$

(2) In Figure 4-2(b) the coordinates of two points P_1 and P_2 are determined as shown. We see that $\Delta x = 6$ and $\Delta y = -12$. The slope is obtained from $\Delta y/\Delta x = -12/6 = -2$. The y-intercept is 24. The equation of this line is obtained by substituting these values into equation (4-1). This gives

$$y = -2x + 24.$$

(3) Figure 4-3 is the graph of some experimental data. It is an example of what a graph should look like for an experiment performed in this laboratory. It shows the distance as a function of the elapsed time. In this graph the elapsed time (on the x-axis) is the independent variable and the distance (on the y-axis) is the dependent variable. The slope is determined from the points P_1 and P_2 on the graph. The calculation of the slope is done on the graph. The y-intercept, b, is shown where the line crosses the y-axis. To find the equation of this line the calculated value for the slope and the value for b shown on the graph are substituted into equation (4 -1). This yields

$$d = 1.47t + 6.5$$

as the equation of the line on the graph. Here t and d, rather than x and y, are used as the variables to represent the elapsed time and the distance traveled.

In drawing a graph of a straight line the student needs to determine how to set up his or her graph, how to draw the best straight line for his or her data and how to analyze this line.

Setting up a Graph

To set up a well drawn graph, the student should:

(a) Use appropriate graph paper that will allow for accurate analysis. Graph paper with 10 divisions per centimeter or 20 divisions per inch is usually a good choice.

(b) Determine a title for the graph so that it would permit quick identification of what the graph is.

(c) Determine the quantity for each axis.

(d) Clearly indicate the units used on each axis.

(e) Determine the scale to be used on each axis. The scales selected should result in a graph that is large enough to allow for easy and accurate analysis. For this laboratory full page graphs on appropriate 8 1/2" x 11" graph paper are generally best.

(f) Be neat.

Example

Figure 4-3 is drawn on graph paper that has 20 divisions per inch (Figure is shown reduced). The scales chosen result in a full page graph with divisions that are easy to work with. The graph has a title, "The Distance as a Function of the Elapsed Time". The x and y axes are clearly labeled to show the quantities and units for each: The x-axis is labeled Elapsed Time (Seconds) and the y-axis Distance (Centimeters).

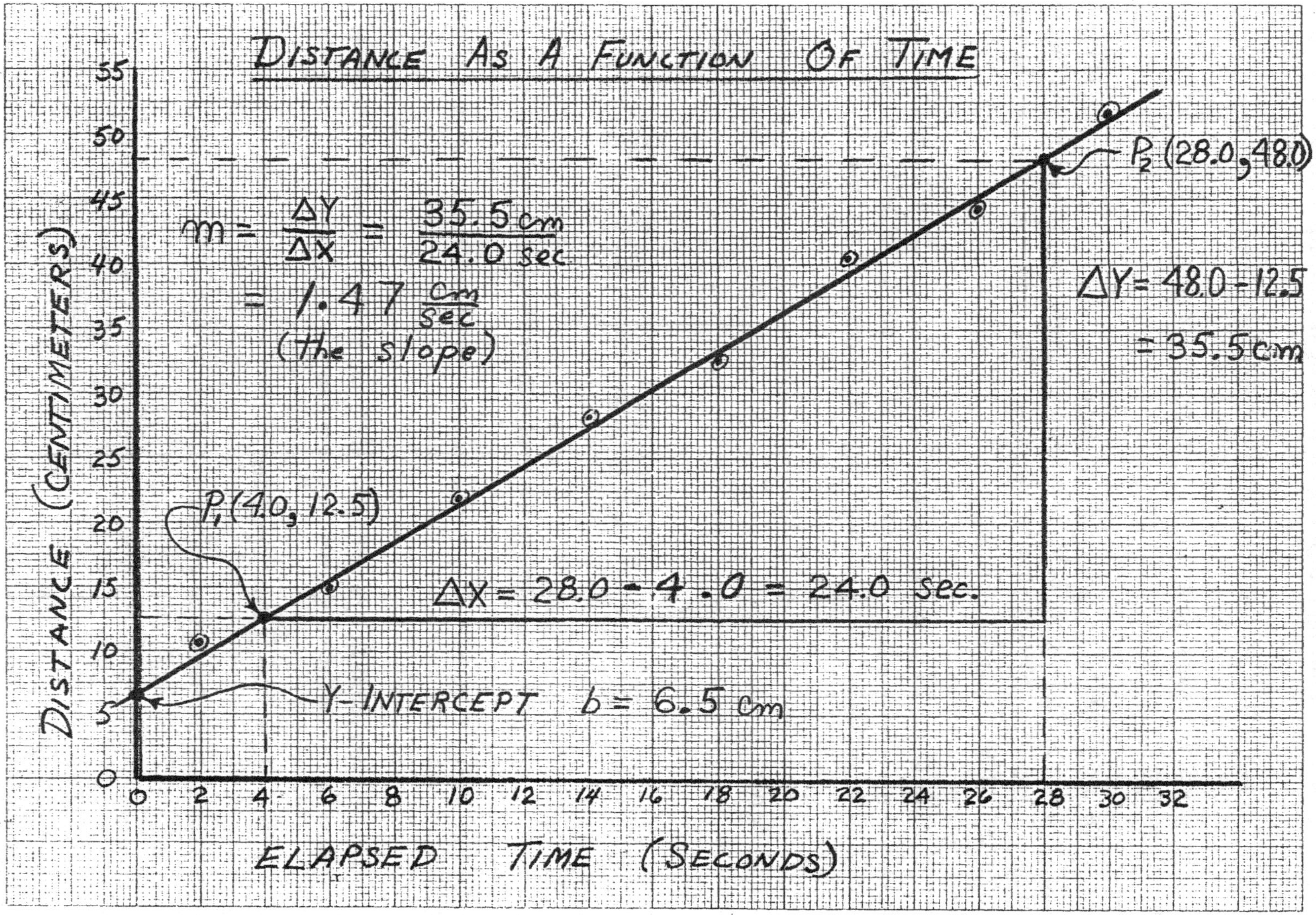

Figure 4-3 A Student Graph

Plotting the Data and Drawing the Straight Line

The student should take care in drawing the graph. He or she should:

(a) Accurately locate each data point at its proper coordinates. Circle each point with a small circle.

(b) Draw the best straight line for the plotted points. This straight line should pass as closely as possible to all points, but does not need to pass through all the points; it might even not pass through any of the points plotted. A straight line once drawn should look like a straight line. Use a straight edge. The line must not have bends or breaks in it.

(c) If any points fall far to either side of the line, then go back to the experiment if possible and check to see if the same results are obtained. Very likely any points that fall far away from the line have a large error associated with them causing the large variation from the pattern followed by the other points.

(d) Be neat.

Example

In the graph in Figure 4-3 all the points are clearly plotted and circled. The best straight line has been carefully drawn so that it is passes as closely as possible to all the points plotted and it has no bends or breaks in it.

The Analysis of a Straight Line

Care should be taken by the student to get accurate results. To analyze a straight line, the student should:

(a) Select two points near opposite ends of the line to use in calculating the slope. Indicate the coordinates of the points selected. Do not use data points.

(b) Draw and calculate Δx and Δy on the graph.

(c) Calculate the slope on the graph from

$$m = \frac{\Delta y}{\Delta x}.$$

Don't forget to include the correct units.

(d) Locate the y-intercept if it is needed.

(e) Write the equation of the line if it is needed.

Example

In Figure 4-3 the slope is determined from the points P_1 and P_2. Neither P_1 nor P_2 are data points. Points P_1 and P_2 are near the ends of the line and their coordinates are clearly indicated. Both Δx and Δy are drawn and calculated on the graph. The slope is calculated on the graph. Its units are clearly shown. The y-intercept, b, is shown where the line crosses the y-axis. The calculated value for the slope and the determined value for the y-intercept are substituted into equation (4-1) to obtain the equation of the line;

$$d = 1.47t + 6.5$$

where t and d are used as the variables to represent the elapsed time and the distance traveled rather than x and y.

Procedure

The distance traveled by an object is determined at intervals of one second as follows:

Elapsed Time (seconds)	Distance Traveled (meters)
1.0	5.2
2.0	7.5
3.0	10.7
4.0	14.4
5.0	17.5
6.0	19.2
7.0	22.6
8.0	26.2
9.0	28.5

You will determine graphically the equation of the line for this data as follows:

1. Construct an x and y coordinate system with labeled axes according to the instructions in *Setting up a Graph* in the discussion of this experiment.

2. Plot the data and draw the best straight line according to the instructions in *Plotting the Data and Drawing the Straight Line* in the discussion.

3. Determine the slope, y-intercept and equation of the line according to the instructions in The Analysis of a Straight line in the discussion.

Questions

1. Real measurements usually have some errors associated with them so that all the data does not necessarily fall on a perfectly straight line. If a student just connects data points he or she will usually get a line with many bends in it. Does such a line have a unique slope? Explain.

2. If you accurately plotted the data from this experiment several times and each time drew the best straight line, do you think that it is possible to get slightly different values for the slope and y-intercept in each case and still to have done everything correctly? Explain.

3. In the discussion you are told to use points near the ends of your line when you calculate the slope. The reason for this is that the effect on the calculated slope of any errors in reading the coordinates of the points are less. To show this do the following:

(a) Assume that the points used to calculate the slope in Figure 14-3 are read as P_1(4.0, 13.0) and P_2(28.0, 47.5). Calculate the slope with these values.

(b) Choose the slope of 1.47 cm/s calculated on Figure 14-3 as the standard value. What is the percent error in the result of part (a).

(c) Calculate the slope of the line in Figure 4-3 using points P_1(12.0, 24.0) and P_2(13.0, 26.0).

(d) Calculate the slope of the line in Figure 4-3 using points P_1(12.0, 24.5) and P_2(13.0, 25.5).

(e) Use the slope calculated in part (c) as the standard. What is the percent error in the slope calculated in part (d).

(f) Notice that in part (a) the values for the y-coordinates of points P_1 and P_2 differed by 0.5 from those on of Figure 4-3. Notice that the y-coordinates of points P_1 and P_2 also differed by 0.5 in part (d) from those in part (c). What effect do these reading errors of 0.5 have on the percent error when the points are taken close together as opposed to far apart.

EXPERIMENT 5

CONVERSION OF INCHES TO CENTIMETERS

Objectives

1. To learn how to convert inches to centimeters.
2. To learn how to plot graphs.
3. To learn how to obtain the slope of a straight line.

Equipment

A meter stick, a yard stick, several items whose dimensions range up to 90cm, graph paper, a straight edge.

Discussion and Theory

The student should review the discussion in Experiment 4, *Linear Graphs*.

In this experiment, we verify that there are 2.54 centimeters per inch. We do this in two ways. In the first method, we make a series of measurements in both centimeters and inches, then determine the conversion for each measurement, and average these values. In the second method, we plot these measurements and determine the conversion from the slope of the line.

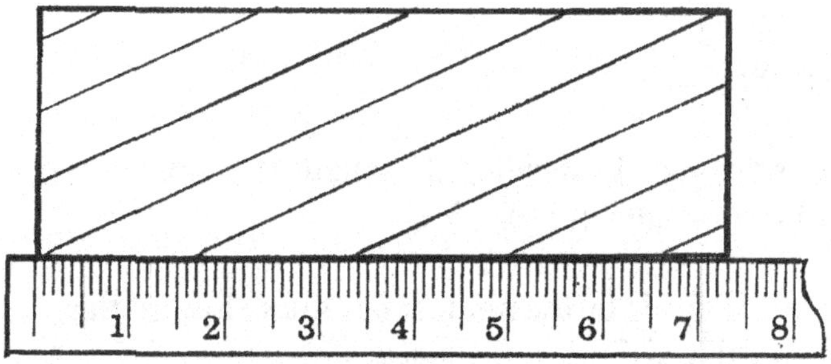

(a) Measuring Length in Centimeters

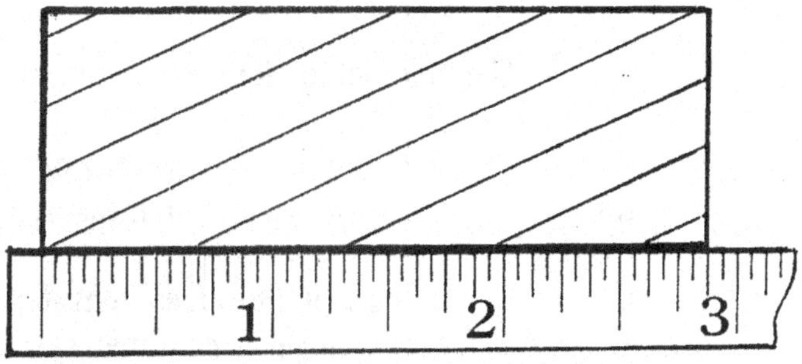

(b) Measuring Length in Inches

Figure 5-1 Measuring Length

In Figure 5-1 an object is measured in centimeters and in inches. In Figure 5-1(a) a meter stick is positioned along the edge to be measured so that the zero falls exactly where the edge of the object starts. The length of the edge of the object is then read at that point along the meter stick where the edge of the object ends at 7.3 cm or 73 mm. If the same object is measured with a yardstick, as shown in Figure 5-1(b), the zero would again have to be positioned at one end of the edge and the ruler would be read as 2 7/8 inches. Most yard sticks are subdivided into inches and fractions of an inch. Sometimes the fractional length is needed as a decimal. The conversion of fractions of an inch to a decimal is demonstrated by means of several examples.

Examples

(1) 6 5/8 inches

$$6 + 5 \div 8 = 6 + 0.625 = 6.625$$

Answer: 6 5/8 inches = 6.625 inches

(2) 29 1/16 inches

$$29 + 1 \div 16 = 29 + 0.0625 = 29.0625$$

Answer: 29 1/16 inches = 29.0625 inches.

The length of 2 7/8 inches which we found in Figure 5-1(b) expressed as a decimal is 2.875 inches. If 7.3 centimeters is divided by 2.875 inches we obtain 2.539... = 2.54 cm/in as expected (within experimental error).

Procedure

1. Use a yard stick and a meter stick to measure in inches and in centimeters the length of 10 objects found in the laboratory. The range of the lengths of the objects measured should vary from 0 to 90 cm. At least three of the objects should have lengths between 0 and 30 cm, at least three between 30 and 60 cm, and at least three between 60 and 90 cm. Record your measurements in Table 5-1. Convert the lengths in inches expressed as fractions to inches expressed as decimals and record.

2. The first method for determining the conversion is as follows:

a. Determine the number of centimeters per inch by dividing the length in centimeters by the length in inches for each item measured. Record your results in Table 5-1.

b. Sum all the values from step 2a above and divide by the number of values. Record this average value for the number of centimeters in an inch in Table 5-1.

c. Find the percent error for your average value of the number of centimeters per inch.

3. The second method for finding the conversion is done graphically.

a. Use the data from Table 5-1 to plot a graph of your values measured in inches along the x-axis and your values measured in centimeters along the y-axis. See *Setting up a Graph* and *Plotting the Data and Drawing the Straight Line* in Experiment 4.

b. Draw the best straight line through these points. See *Plotting the Data and Drawing the Straight Line* in Experiment 4.

c. Determine the slope of your best straight line as follows (See *The Analysis of a Straight Line* in Experiment 4):

-1- Choose a point near the bottom of your line. This point must actually be on your line. *Don't just use a data point!* Call this point P_1. Determine the x and y coordinates for point P_1. Record the value of the coordinates on your graph in the form $P_1(x_1,y_1)$.

-2- Choose a point near the top of your line. This point must actually be a point on your line. *Don't just use a data point!* Call this point P_2. Determine the x and y coordinates for point P_2. Record the coordinates on your graph in the form $P_2(x_2,y_2)$.

-3- Find the change in the x-coordinate by subtracting x_1 from x_2. Call this Δx. Find the change in the y-coordinates by subtracting y_1 from y_2. Call this Δy. Draw Δx and Δy on your graph. Show the calculations for Δx and Δy on the graph.

-4- The number of centimeters in an inch is obtained from the slope of the line which is Δy divided by Δx. Determine the slope of your line. Show this calculation on your graph.

d. Determine the percent error. Show your calculation on your graph.

Questions

1. The equation of a straight line is of the form $y = mx + b$. (a) For the graph you drew, what did the variable x represent? (b) The variable y? (c) What were the values for m and b for your graph? (d) If there were absolutely no error in your data and your best straight line, what would your values of m and b have been?

2. If several different people drew their own graph of your data, would you expect all of them to obtain exactly the same value for the slope as you obtained? Explain.

3. Convert the following to centimeters: (a) one foot, (b) one yard, (c) 1.53 inches, (d) 23.6 inches, (e) 0.000814 inches.

4. Convert the following to inches: (a) one millimeter, (b) one meter, (c) one kilometer, (d) 12.7 centimeters, (d) 0.0418 centimeters

EXPERIMENT 6

THE SIMPLE PENDULUM

Objectives

1. To study how the period of a simple pendulum is affected (a) when the length of the pendulum is changed and (b) when the mass of the pendulum is changed.

2. To see how we can analyze the equation of the simple pendulum to determine the acceleration due to gravity.

Equipment

Stop watch, stand, cord, three spherical masses, meter stick.

Discussion and Theory

A *simple pendulum* consists of a mass suspended from a cord, where the mass is permitted to swing back and forth freely through a small angle as shown in Figure 6-1. The time it takes the pendulum to swing from one side to the other and back to its starting position is called the *period.* A simple pendulum of length L has a period of

$$T = 2\pi\sqrt{\frac{L}{g}} \tag{6-1}$$

where g is the acceleration due to gravity, 9.8 m/s^2. Notice that the period of the simple pendulum depends only on its length and the acceleration due to gravity. The period does not depend on the mass of the pendulum.

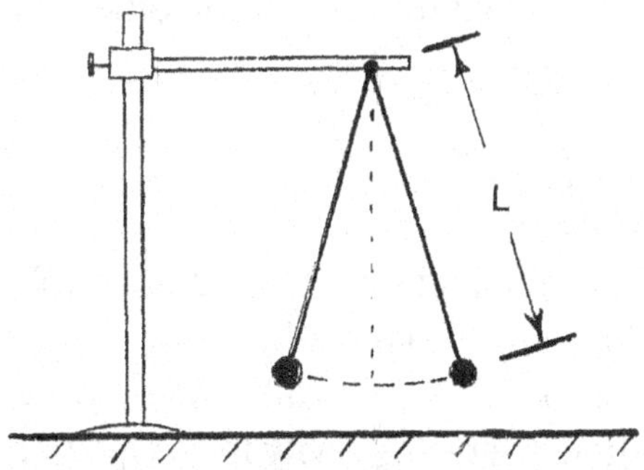

Figure 6-1 A Simple Pendulum

Equation (6-1) is not a straight line; it is not of the form $y = mx + b$. When we square both sides of equation (6-1) we obtain

$$T^2 = (4\pi^2 / g)\, L. \qquad (6\text{-}2)$$

If we let $y = T^2$, $x = L$, and the slope $m = (4\pi^2 / g)$, we obtain

$$y = m\,x \qquad (6\text{-}3)$$

for equation (6-2). Equation (6-3) is the equation of a straight line with a y-intercept at zero ($b = 0$). From the expression for the slope, m, we can solve for the acceleration due to gravity

$$g = 4\pi^2 / m. \qquad (6\text{-}4)$$

Procedure

1. Determine the mass of one of the spherical masses on a balance and record in Table 6-1

2. With this mass construct a simple pendulum with a length of 0.20 meters. The length of the pendulum is measured from the top secured end of the string to the center of the hanging mass as shown in Figure 6-1.

3. Pull the pendulum to one side so that it makes a small angle with the vertical. One complete oscillation of the pendulum goes from one side to the other and back to its starting position. Release the pendulum and with a stopwatch determine the time for ten complete oscillations. Record this time in Table 6-1.

4. Divide the time determined in step 3 by ten to obtain the period, T. Record this period in Table 6-1.

5. Repeat the procedure in steps 2 through 4 four more times; each time with a different length for the simple pendulum. Use the lengths 0.40 m, 0.60 m, 0.80 m and 1.00 m.

6. Repeat the procedure in steps 1 through 5 for each of the other masses you are given. Use Tables 6-2 and 6-3 for your data.

7. Analyze the data for each mass as follows:
   (a) Square all the periods and record.
   (b) Construct a graph of your data. Plot L on the x -axis and T^2 on the y-axis.
   (c) Draw the best straight line for your points.
   (d) Calculate the slope, m, of your line. Show all work on your graph.
   (e) Determine the acceleration, g, due to gravity from equation (6-4). Record your result.
   (f) Determine the percent error in your acceleration due to gravity. The standard value for the

acceleration due to gravity is 9.8 m/s^2.

8. Answer the question at the end of the data sheet.

Questions

1. To find the period of each pendulum we measured the time for ten oscillations and then divided the result by ten. Why is this better then trying to measure the time of just one oscillation?

2. When the pendulum of a grandfather's clock takes too long to make each oscillation the clock runs slow. How should the length of the pendulum be adjusted to make the clock run faster?

Name____________________ Date Performed____________________

Lab Partners____________________ Date Submitted____________________

Table 6-1

Pendulum with First Mass - Variation of Period with Length

Mass of pendulum: M_1 = __________

Length of Pendulum (m)	Time for Ten Oscillations (s)	Period (s)	$(\text{Period})^2$ (s^2)
0.20			
0.40			
0.60			
0.80			
1.00			

slope = __________ experimental g = __________ % error = __________

Table 6-2

Pendulum with Second Mass - Variation of Period with Length

Mass of pendulum: M_2 = __________

Length of Pendulum (m)	Time for Ten Oscillations (s)	Period (s)	$(\text{Period})^2$ (s^2)
0.20			
0.40			
0.60			
0.80			
1.00			

slope = __________ experimental g = __________ % error = __________

Name________________________________

Table 6-3

Pendulum with Third Mass - Variation of Period with Length

Mass of pendulum: M_3 = ____________

Length of Pendulum (m)	Time for Ten Oscillations (s)	Period (s)	$(\text{Period})^2$ (s^2)
0.20	____________	____________	____________
0.40	____________	____________	____________
0.60	____________	____________	____________
0.80	____________	____________	____________
1.00	____________	____________	____________

slope = ____________ experimental g = ____________ % error = ____________

Compare the periods obtained for L = 0.20 cm for each mass. Does it appear as if the mass has an effect on the period? Do the same for each of the other lengths. State your observations.

____

____

____

____

____

____

____

____

EXPERIMENT 7

DENSITIES

Objectives

1. To learn how to use a meter stick, a vernier caliper and a micrometer.
2. To learn how to use an electronic balance.
3. To determine the densities of several substances.

Equipment

A meter stick, a vernier caliper, a micrometer, an electronic balance and several samples whose densities will be determined.

Discussion and Theory

The student is to review the discussion of Experiment 2, *Understanding Errors and Units*, prior to performing this experiment.

The Electronic Balance

The mass of an object on the platform of an electronic balance is shown in the display (See Figure 7-1). Electronic balances have a tare adjustment. The tare adjustment is used to set the scale to read zero.

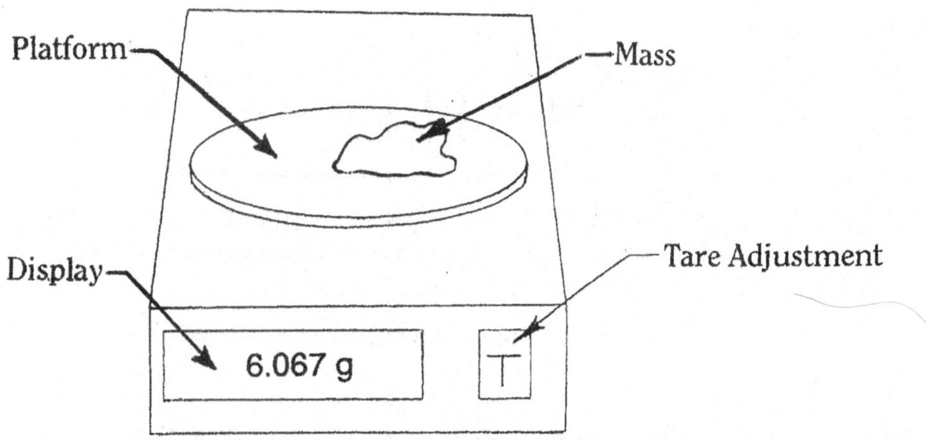

Figure 7-1 An Electronic Balance

Verniers and Micrometers

When it is necessary to make a more accurate measurement than is possible with just a ruler either a vernier or a micrometer should be used. The degree of accuracy required will determine the instrument to be used.

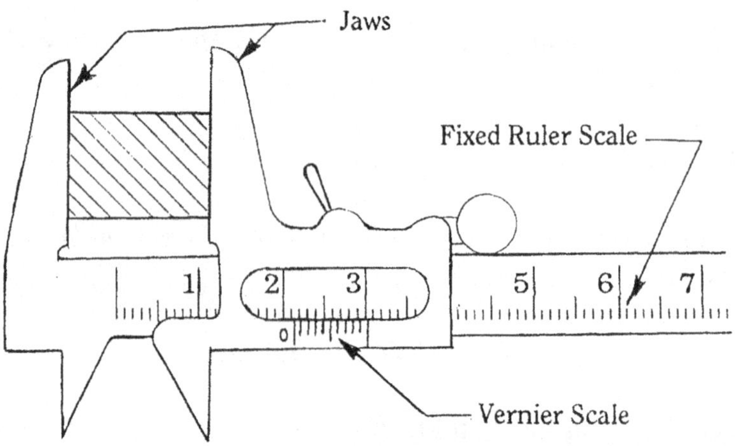

(a) A Vernier Caliper Measuring a Small Block

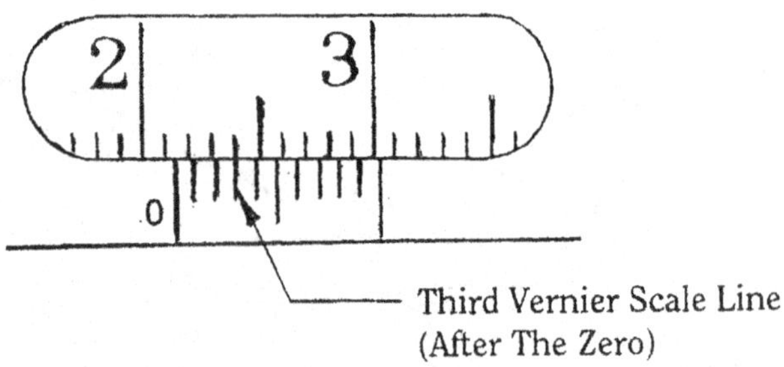

(b) Reading The Vernier Scale

Figure 7-2 Making a Measurement with a Vernier

The Vernier

When a vernier caliper is used (See Figure 7-2), the jaws of the caliper are first closed on to the object to be measured. The fixed ruler scale is then read against the zero on the vernier scale. If we assume the fixed ruler scale is marked in centimeters with subdivisions of 0.1 cm, then this scale is read to the nearest 0.1 centimeter. The vernier scale is used to find the next digit in the measurement. To do this one looks along the vernier scale until one finds the vernier line that lines up with a line on the fixed ruler scale. This vernier line gives the last digit.

Example

For the vernier in Figure 7-2 the reading of the fixed ruler scale is 2.1 cm. To obtain the next digit the vernier scale is used [see Figure 7-2(b)]. The 3rd line from the zero on the vernier scale lines up with a line on the fixed ruler scale. This indicates the reading is 0.03 cm. Thus, the total reading for this measurement is

$$(2.1 + 0.03) \text{ cm} = 2.13 \text{ cm}.$$

The Micrometer

If even more accurate measurements are needed than can be obtained with a vernier, a micrometer should be used (See Figure 7-3). The object to be measured is placed against the anvil of the micrometer. The thimble is then slowly turned by means of the thumbscrew until the bolt closes on the object. This is continued slowly until the thumbscrew starts to slip. In a micrometer with a fixed scale marked in millimeters the thimble makes two revolutions per millimeter. For each revolution the thimble makes it moves only 0.5 mm along the fixed scale. The fixed scale is therefore marked at each half millimeter. In order to read the micrometer, read the fixed scale first to the nearest half millimeter then the micrometer scale and add the micrometer scale reading to the fixed scale reading.

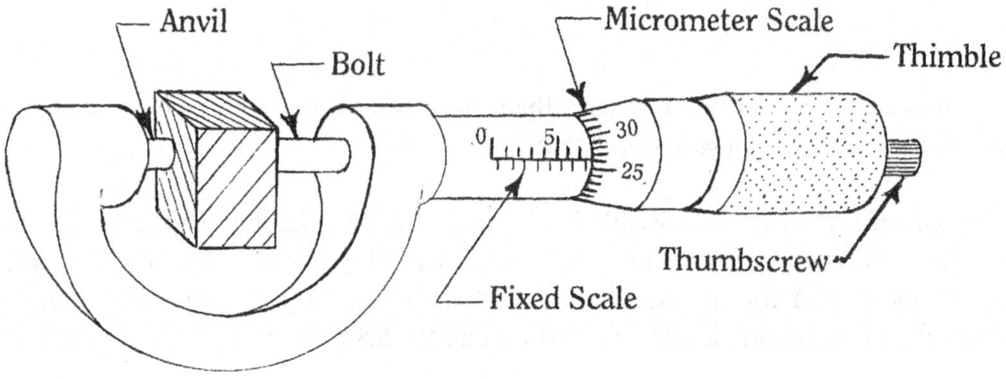

Figure 7-3 Making a Measurement with a Micrometer

Example

In the micrometer shown in Figure 7-3, half divisions are shown along the bottom of the fixed millimeter scale. The fixed scale reads 7.50 mm and the micrometer scale reads .26 mm. The total reading is

$$(7.50 + .26) \text{ mm} = 7.76 \text{ mm}.$$

Whenever a student starts to make a measurement with a micrometer, he or she should check that when the instrument is fully closed by means of the thumbscrew it actually reads zero. If it does not, it should be determined by how much it reads above or below zero. This is called the zero reading. If a fully closed micrometer reads +0.02 mm, it will read 0.02 mm more than the true value (zero reading = +0.02). Therefore, the student must subtract 0.02 mm from any measurement made with this micrometer. If on the other hand, the micrometer reads 0.02 mm in the other direction (zero reading = -0.02) then 0.02 mm must be added to any reading.

Example

If in the example in Figure 7-3, the micrometer read +0.02 mm when fully closed, then the final reading should be corrected by subtracting 0.02 mm. Therefore, the corrected reading is

$$(7.76 - 0.02) \text{ mm} = 7.74 \text{ mm}.$$

Density

The density of an object is its mass per unit volume, i.e.,

$$\text{Density} = \frac{\text{Mass}}{\text{Volume}} \tag{7-1}$$

If the mass is in grams and the volume in cm^3, then the density has units of g/cm^3. Table 7-1 lists the standard values for the densities of various substances.

In this experiment, the student will be given several samples for which the density is to be determined. For each sample, the mass is determined by placing the object on the electronic balance and reading the mass from the display of the balance. The volume is obtained by measuring the necessary dimensions of each sample and substituting into the appropriate formula. Figure 7-4 indicates

Table 7-1
The Densities of Various Substances

Substance	Density (g/cm^3)
Aluminum	2.70
Brass	8.5
Copper	8.92
Iron	7.86
Lead	11.3
Steel	7.7
Tin	7.29
Wood: Oak	.6 - .9
Wood: Pine	.35 - .85
Zinc	7.13

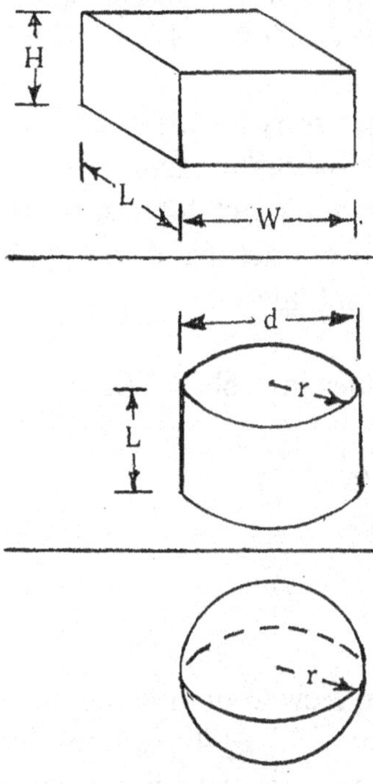

Volume = $L \times W \times H$

Volume = $\pi r^2 L$

$= \frac{1}{4} \pi d^2 L$

Volume = $\frac{4}{3} \pi r^3$

Figure 7-4 Formulas for Determining Volumes

the needed formulas for finding the volumes of various common shapes. The volume for each sample is determined three times: first with measurements made with a meter stick, then with a vernier caliper, and finally with a micrometer. Each of these instruments has a different precision. The meter stick results in measurements accurate to the nearest millimeter or 0.1 centimeter. The vernier caliper increases the precision of the measurements to the nearest 0.1 millimeter or 0.01 centimeter. Finally, the micrometer increases the precision to the nearest 0.01 millimeter or 0.001 centimeter.

Since the measuring instruments have different precisions, then the calculated results will have different numbers of significant figures. The volumes and densities calculated from the micrometer measurements have the greatest number of significant figures. The volumes and densities calculated from the meter stick measurements have the least number of significant figures.

Procedure

The student will be given several samples for which the densities will be determined For each sample the student should do the following:

1. With the electronic balance determine the mass of the sample and record its value in Table 7-2. Be certain to record the value to the proper number of significant figures and to include units.

2. Measure each of the physical dimensions necessary for the calculation of the volume of each object: (a) with the meter stick, (b) with the vernier, (c) with the micrometer. All measurements should be recorded in centimeters to the proper number of significant figures. Record these values in the appropriate spaces in Table 7-2. Leave spaces blank for quantities that do not apply. Calculate the density for each set of readings to the proper number of significant figures.

3. Compare your results with the standard values in Table 7-1 Calculate and record the percent error.

Questions

1. In this experiment we used an electronic balance. (a) How would the number of significant figures of the mass be affected if the balance used had a precision of 0.1 g? (b) How would it affect the number of significant figures calculated for the values of the density for your first sample when the lengths are measured with the meter stick, the vernier and the micrometer?

2. Of all the measuring instruments used in this experiment, which would you use to determine the thickness of a piece of paper.

3. Which of the following is more difficult to determine accurately with a ruler, the width of a rectangle or the diameter of a sphere.

EXPERIMENT 8

PROPERTIES OF RIGHT TRIANGLES AND THE TRIGONOMETRIC FUNCTIONS

Objectives

1.To study some of the properties of right triangles
2. To learn the trigonometric functions.

Equipment

A protractor, paper and a ruler.

Discussion and Theory

A right triangle is any triangle with one of its angles equal to 90°. The longest side of such a triangle is opposite the 90° angle and is called the *hypotenuse*. The remaining two sides are called *legs*. On the right triangle ABC in Figure 8-1, side c is the hypotenuse and sides a and b are the legs. One of the best known properties of right triangles is the *Pythagorean Theorem*. This states: *In a right triangle the square of the hypotenuse is equal to the sum of the squares of the legs*. For triangle ABC

$$c^2 = a^2 + b^2. \tag{8-1}$$

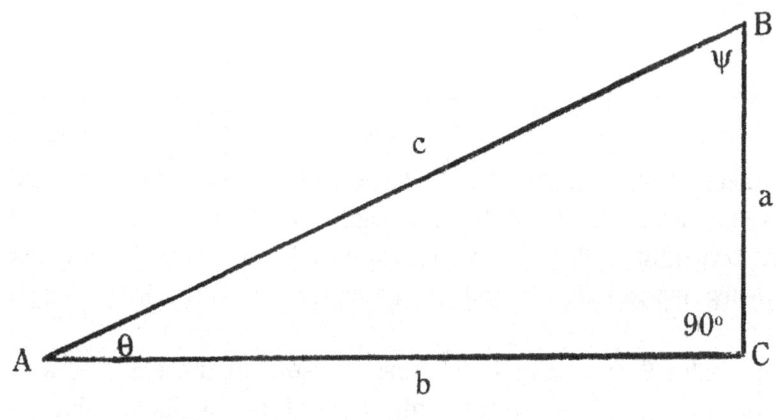

Figure 8-1 A Right Triangle

The sum of the angles of any triangle is 180°. Since in a right triangle one of the angles is 90°, then the sum of the remaining two angles must be 90°.

In right triangle ABC in Figure 8-1 the vertex of angle θ is at point A. The leg a that the angle θ faces is called its *opposite leg*. The leg b that forms one side of the angle θ is called its *adjacent leg*. The angle ψ with its vertex at point B has legs b and a as its opposite and adjacent legs respectively.

Three trigonometric functions are called the sine (sin), cosine (cos) and tangent (tan) functions. For the angle θ they are defined as follows:

$$\sin(\theta) = \frac{\text{Opposite Leg}}{\text{Hypotenuse}} = \frac{a}{c} \tag{8-2}$$

$$\cos(\theta) = \frac{\text{Adjacent Leg}}{\text{Hypotenuse}} = \frac{b}{c} \tag{8-3}$$

$$\tan(\theta) = \frac{\text{Opposite Leg}}{\text{Adjacent Leg}} = \frac{a}{b} \tag{8-4}$$

For the angle ψ,

$$\sin(\psi) = b/c, \qquad \cos(\psi) = a/c\,, \qquad \tan(\psi) = b/a.$$

Procedure

1. On a clean piece of paper neatly draw a large right triangle ABC with the vertex of its right angle at point C. Make certain all sides are drawn straight and that all sides meet at the corners of the triangle. Label the two non-right angles with vertices at points A and B as θ and ψ respectively. Label the right angle λ. Label the leg opposite θ as a and the leg opposite ψ as b. Label the hypotenuse as c.

2. Measure the angles θ, ψ and λ. Determine the sum of these angles and compare with the standard value of 180° by calculating the percent error. Record the results in Table 8-1.

3. Measure and record in Table 8-2 the lengths a, b, and c.

4. Calculate c using the measured lengths of the legs a and b from step 3, and the Pythagorean Theorem. From equation (8-1)

$$c = \sqrt{a^2 + b^2}.$$

Calculate the percent error in this calculated value for c. Use the measured value from step 3 as the standard value. Record your result in Table 8-2.

5. Calculate sin(θ), cos(θ), tan(θ), sin(ψ), cos(ψ) and tan(ψ) from the angles θ and ψ measured in step 2 and record in Table 8-3.

6. Calculate and record in Table 8-3, sin(θ), cos(θ), tan(θ), sin(ψ), cos(ψ) and tan(ψ) from the measured lengths of sides a, b and c from step 3 and the definitions of the trigonometric functions.

7. Compare the experimental results from step 6 with those from step 5 by calculating the percent error of each result in step 6. Use the values calculated from the angles in step 5 as the standard values. Record your results in Table 8-3.

Questions

1. If you drew a right triangle with legs exactly half the length of those in triangle ABC, what values do you think you would get for the trigonometric functions for this triangle?

2. A certain right triangle has one angle equal to 53.1° and its are legs equal to 6.0 m and 8.0 m. (a) What is the other non-right angle? (b) What is the length of the hypotenuse?

3. At what angles does (a) sin(θ) = 1, (b) cos(θ) = 1, (c) tan(θ) = 1?

4. At what angle does (a) sin(θ) = 0, (b) cos(θ) = 0, (c) tan(θ) = 0?

5. Which trigonometric function has values greater than one for angles greater then 45°?

EXPERIMENT 9

LAW OF COSINES

Objectives

1. To verify that the sum of the angles of a triangle is 180°.
2. To verify the Law of Cosines.

Equipment

A protractor, paper, a ruler.

Discussion and Theory

The student should be aware that the sum of the angles of any triangle is 180°. This can be verified directly by measuring the angles with a protractor and adding the measured angles.

The general relationship among the sides of any triangle is given by the *Law of Cosines*. This states that

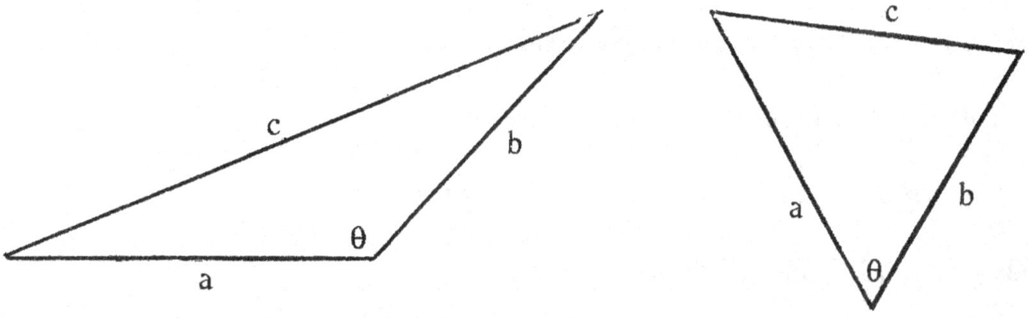

(a) An Obtuse Triangle (one angle is greater than 90°)

(b) An Acute Triange (all angles less than 90°)

Figure 9-1 Labeling of the Sides of Triangles for the Law of Cosines (The angle θ is always taken opposite the side chosen as c)

$$c^2 = a^2 + b^2 - 2ab\cos(\theta) \qquad (9\text{-}1)$$

where a, b, and c are the sides of the triangle and θ is the angle formed by sides a and b and is opposite side c. (See Figures 9-1(a) and (b)).

Cos(θ) is positive for θ less than 90° and is negative for values of θ greater than 90° and less than 180°. At θ = 90°, cos(θ) = 0 and the triangle is a right triangle. In this case equation (9-1) becomes the familiar Pythagorean Theorem

$$c^2 = a^2 + b^2 \qquad (9\text{-}2)$$

where c is the hypotenuse and a and b are the legs of the triangle.

In this experiment, to verify the Law of Cosines, you will draw a triangle and label its sides a, b, and c. You will measure the lengths of all the sides of your triangle and the angle opposite side c (the angle θ). The length of side c will then be calculated from

$$c = \sqrt{a^2 + b^2 - 2ab\cos(\theta)} \qquad (9\text{-}3)$$

and compared to your measured length for side c.

Procedure

1. Use a straight edge and draw a large triangle on a piece of 8 1/2" X 11" paper so that each of the sides is a different length and none of the angles is 90°.

2. Label the sides a, b, and c. Label the angle opposite side c as θ.

3. Measure each of the angles. Record these values in Table 9-1.

4. Find the sum of these angles and record this value in Table 9-1.

5. Compare this sum with 180° and find the percent error.

6. Measure the length of each of the sides of your triangle. Record these values in Table 9-2. Remember to include units.

7. Calculate a^2. Record in Table 9.2.

8. Calculate b^2. Record in Table 9.2.

9. Calculate 2ab cos(θ). Record in Table 9.2. Indicate if the value of 2ab cos(θ) is positive or negative. Remember to include units.

10. From the values obtained in steps 7, 8, and 9 calculate $a^2 + b^2 - 2ab\cos(\theta)$. Pay attention to the signs. Record this value in Table 9-2.

11. Take the square root of the result found in step 10. Compare this value with your measured length for side c, the standard value. Determine the percent error and record in Table 9-2.

Questions

1. A triangle has sides a = 3.0 m and b = 4.0 m . What is the length of the third side c if the angle between sides a and b is (a) 50°? (b) 90°? (c) 140°?

2. When θ = 180°, a and b become coincident to c (i.e., a and b form a straight line that falls over c). To verify this do the following: (a) choose any lengths you wish for a and b, (b) calculate the length c from the law of cosines, (c) calculate the length of a + b. What do you observe about the value of c and the value of a + b?

3. A triangle has sides a = 5.0 m and b = 4.0 m. At what angle θ will c = 9.0 m. Is c actually the side of a triangle? Explain.

4. Show that if sides a and b are each doubled and the angle θ remains the same, then the length of side c is also doubled (Hint: Consider a triangle with sides A, B, and C. Let a = 2A and b = 2B in equation (9-1)).

Name________________________________ Date Performed____________________

Date Submitted____________________

Table 9-1

Sum of the Angles of a Triangle

Angle opposite side a = ____________________

Angle opposite side b = ____________________

Angle opposite side c (angle θ) = ______________

Sum of the angles (experimental) = ______________

Sum of the angles (theoretical) = ___ 180° ________

Percent Error = __________________________

Table 9-2

Law of Cosines

a = ____________________	a^2 = ________________
b = ____________________	b^2 = ________________
c = ____________________	XXXXXXXXXXXXXXXXXXXXXXX
Angle θ = ______________	$2ab\cos(\theta)$ = ____________
$a^2 + b^2 - 2ab\cos(\theta)$ = ________	XXXXXXXXXXXXXXXXXXXXXXX
$\sqrt{a^2 + b^2 - 2ab\cos(\theta)}$ = ______	XXXXXXXXXXXXXXXXXXXXXXX
Percent Error = ____________	XXXXXXXXXXXXXXXXXXXXXXX

EXPERIMENT 10

THE CENTER OF MASS

Objective

To find the center of mass of an object.

Equipment

Manila folder, cord, small weight, stand, hole puncher, scissors, paper clip.

Discussion and Theory

In this experiment, we will find the center of mass of several shapes. A physical object can be shown to behave as if all its mass is concentrated at one point. This point is called the *center of mass.* To find the center of mass, we will use the property that, whenever an object hangs freely, the center of mass will fall directly below the suspension point, provided that the suspension point is not the center of mass.

If an object is suspended so that it hangs freely from a point A, as shown in Figure 10-1(a), the center of mass must fall directly below point A. If a vertical line is drawn through A, the center of mass must be on that line. To determine the actual point where the center of mass is located, the object must be suspended from a second point B and a vertical line drawn through point B, as shown in Figure 10-1(b). The intersection of the two lines will be at the center of mass (c.m.), since it is the only point common to both lines. To check the accuracy of this determination, the object should be suspended from a third point C, as shown in Figure 10-1(c). The center of mass will be on the vertical through C at the intersection of the lines previously drawn through points A and B.

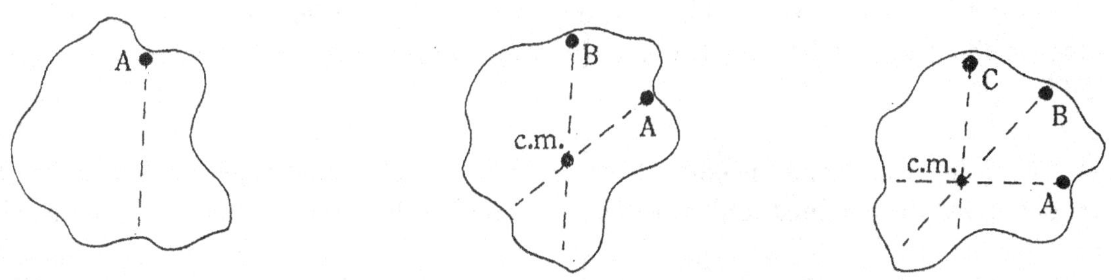

(a) Object suspended from first point

(b) Object suspended from second point

(c) Object suspended from third point

Figure 10-1 Determination of the Center of Gravity

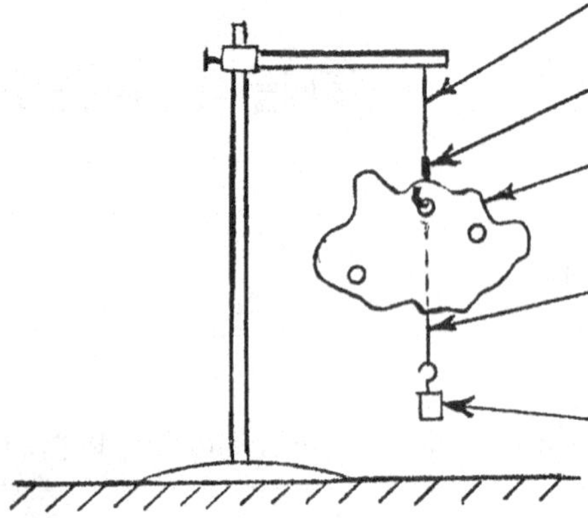

Figure 10-2 Experimental Set-Up to Determine the Center of Gravity

Procedure

Draw and cut out a triangle and an irregular shaped object from a manila folder. With a hole punch, punch out three holes at three different locations near the periphery of each object. Label these holes A, B, and C. Determine the location of the center of mass of each of these objects as follows:

1. Bend a paper clip to form a hook. Attach the hook to one end of a piece of cord and suspend the cord at the other end from a stand. Attach a small weight to the end of a second piece of cord. Suspend the other end of the second cord from the hook. See Figure 10-2.

2. Suspend the object whose center of mass is to be determined from the hook at hole A as shown in Figure 10-2. Draw a vertical line down from the center of A along the weighted cord.

3. Suspend the object from the hook at hole B and draw a vertical line down from the center of B along the weighted cord. Where the two lines from A and B intersect is the center of mass. Mark this point "c.m.".

4. As a check , suspend the object from the hook at hole C and draw the vertical line through the center of C along the weighted cord. Does this line cross at the intersection of the other two lines?

Question

1. When a certain rectangle is suspended freely from any of its corners it is observed that one of is diagonals is always vertical. Where is the center of mass of the rectangle?

EXPERIMENT 11

INDIRECT MEASUREMENTS OF AREAS

Objective

To determine areas indirectly

Equipment

A protractor, a manila folder, an electronic balance, a meter stick, scissors.

Discussion and Theory

Sometimes physical restrictions make it difficult, if not impossible, to measure a particular quantity directly. Consequently, a scientist is forced to devise an indirect method for making the measurement. In this experiment we will measure areas indirectly.

If a two dimensional object is drawn on a piece of a manila folder and then cut out, the area, A_x, of the cut-out shape is proportional to the mass, M_x, of the cut-out. If M_x can be found, the area A_x of the cut-out shape can now easily be determined by using proportions, provided that the mass, M_s, and area, A_s, of another shape cut out from the same manila folder are known. Therefore,

$$\frac{A_x}{A_s} = \frac{M_x}{M_s} \quad (11\text{-}1)$$

If we solve for A_x, we obtain

$$A_x = \frac{M_x (A_s)}{M_s} \quad (11\text{-}2)$$

We will verify equation (11-2) by calculating the area of a triangle directly from

$$\text{Area of a triangle} = \frac{1}{2}(\text{base}) \times (\text{altitude}) \quad (11\text{-}3)$$

and then indirectly from equation (11-2). If these results agree, we can then feel confident to apply the indirect method to shapes whose areas cannot easily be determined from direct measurements.

Procedure

1. Use a straight edge and draw a large triangle on one half of a manila folder. Save the other half of the folder.

2. *Calculation of the area of the triangle from direct measurements:*

a. Choose any side of the triangle as the base. Measure the base and record this value in Table 11-1.

b. Drop a perpendicular from the vertex of the angle opposite the base to the base (Use a protractor). This perpendicular is the altitude.

c. Measure the altitude of the triangle. Record this value in Table 11 -1.

d. Calculate the area of this triangle from equation (11-3). Record this value in Table 11-1.

3. *Calculation of the area of the triangle from indirect measurements:*

a. Carefully draw a square 10.0 cm by 10.0 cm on the unused part of the manila folder.

b. Calculate the area of this square. Record this value in Table 11 -2.

c. Carefully cut out this square. Be certain to cut on the lines!

d. Use the electronic balance and determine the mass of this square, M_s. Record this value in Table 11-2.

e. Carefully cut out the triangle drawn in step 1. Be certain to cut on the lines!

f. Determine the mass, M_t, of your triangle and record this value in Table 11-2.

g. Determine the area, A_t, of your triangle from equation (11-2) (A_t is the A_x of equation (11-2)). Record this value in Table 11-2.

h. Find the percent error in this area. Use the value determined from direct measurements in part 2 as your standard value.

4. *Calculation of the area of an irregular figure from indirect measurements:*

a. Cut out an irregular figure from the remaining manila folder. Make this as large as possible.

b. Use the electronic balance to determine the mass, M_i, of this irregular figure. Record M_i in Table 11-2.

c. Follow the same procedure used in step 3 for the area of the triangle to determine the area A_i of your irregular figure from equation (11-2) (A_i is the A_x of equation (11-2). Record your results in Table 11-2.

Questions

1. Which of the following do you think cannot be measured directly. Why? (a) The distance of the moon from the earth, (b) The size of a hydrogen atom, (c) The height of a tree, (d) The depth of the Atlantic Ocean at some point half way between the US and Europe, (e) The temperature of a star.

2. You have a package of 500 sheets of paper. How could you determine the thickness of one sheet with only a ruler?

3. Based on your results for the triangle do you have confidence that the area of your irregular shape is reasonably close to the actual area. Explain.

Name________________________________ Date Performed________________________________
Date Submitted________________________________

Table 11-1
Area of a Triangle from Direct Measurements

Base________________________________

Altitude________________________________

Area of triangle (A_t)________________________________

Table 11-2
Areas from Indirect Measurements

Area of square (A_s)________________________________

Mass of square (M_s)________________________________

Mass of triangle (M_t)________________________________

Area of triangle from equation (11-2)________________________________

Area of triangle from Table 11-1________________________________

% error________________________________

Mass of irregular figure (M_i)________________________________

Area of irregular figure (A_i)________________________________

EXPERIMENT 12

INDIRECT MEASUREMENTS OF HEIGHTS

Objective

To measure heights indirectly.

Equipment

A straw, a protractor with a hole at the vertex, cord, small weight, meter stick or a metric tape measure, scotch tape.

Discussion and Theory

In this experiment, we will measure the height of several objects with the device shown in Figure 12-1. It is made from a protractor that has a hole in it at the vertex. A straw is secured with tape along the straight edge of the protractor and a weighted cord is passed through the hole and fastened in such a way that the weighted end of the cord can hang freely a short distance below the angle scale of the protractor.

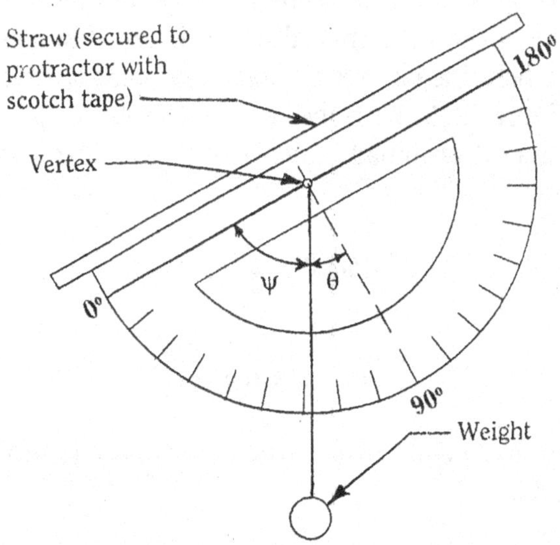

ψ = Reading on protractor scale
θ = Angle of elevation = $90^{\circ} - \psi$

Figure 12-1 The Angle Measuring Device

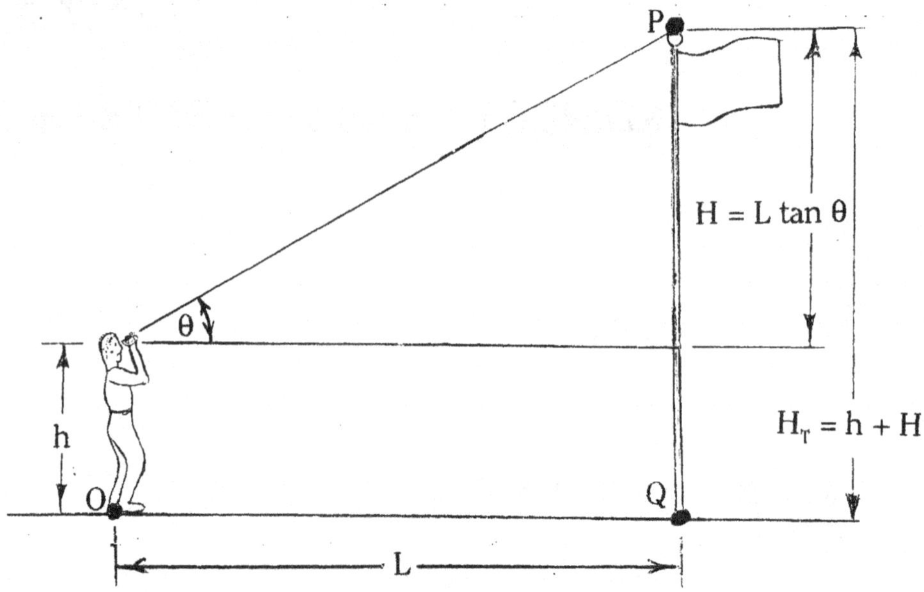

Figure 12-2 Use of the Angle Measuring Device to Measure Height

To measure the height of an object, one first sights through the straw until one sees the top of the object at point P, as shown in Figure 12-2. The angle of elevation, θ, is the angle between the 90° mark on the protractor and the weighted cord, as shown in Figure 12-1. The angle, ψ, is the angle indicated on the protractor scale by the weighted cord, as shown in Figure 12-1. The angle of elevation, θ, is calculated from $\theta = 90 - \psi$. The horizontal distance L from the observer to the object is then measured on the ground from where the observer is standing (point O) to directly under the top of the object (point Q), as shown in Figure 12-2. With the angle of elevation, θ, and the length L both known, the height above eye level, H, can be determined as shown in Figure 12-2 from

$$\tan\theta = \frac{H}{L} \tag{12-1}$$

$$H = L\tan\theta \tag{12-2}$$

To obtain the total height H_T, we must add the distance, h, from the ground to the eye of the observer. The total height, H_T, of the object is:

$$H_T = h + H \tag{12-3}$$

$$= h + L\tan\theta \tag{12-4}$$

Procedure:

Two people are needed for this experiment, one to sight through the measuring device and the other to read the angle. Your instructor will tell you which objects on campus to measure (Don't forget all measured quantities must be recorded with units!!!). The procedure is as follows:

1. Measure the height h from the ground to the eye of the person using the protractor. Record the height, h, of the eye of this person in Table 12-1.

2. Sight through the straw to the top of the object being measured. Read the angle ψ which the weighted cord makes with the protractor from 0° and record in Table 12-1.

3. Calculate the angle of elevation θ from $\theta = 90^\circ - \psi$.

4. Measure the distance L of the object from the observer and record in Table 12-1.

5. Calculate the total height of the object, H_T, from equation (12-4) and record in Table 12-1.

6. The student is to determine the height of the object two more times at different distances, L. If done properly, the three values obtained for the total height, H_T, should be similar.

7. Determine the average value for the total height.

8. For each individual value of the total height, determine the percent deviation from the average value of the total height. The percent deviation is calculated from

$$\%\ \text{deviation} = \frac{|\, H_T - (\text{average value of } H_T)\, |}{\text{average value of } H_T} \times 100$$

9. If possible, determine the height of the object measured above either by direct measurement or by some other method (e.g. For a brick building measure the height of ten bricks and count the number of bricks to the top. Determine the height of one brick and multiply by the number of bricks). Use this result as your standard value and determine the percent error for each height H_T calculated above.

10. Repeat the above for as many objects as required.

Questions

1. If you had to determine the heights of various objects in your neighborhood, for what objects would you use the indirect measurement of height method used in this experiment?

2. For the objects chosen in question 1, what would be involved in measuring them directly?

EXPERIMENT 13

KINEMATICS - CONSTANT VELOCITY

Objective

To study constant velocity

Equipment

Air track, glider with an 8 inch long cord secured at its top, blower, spark timer, masking or scotch tape, rubber band.

Discussion and Theory

The average velocity, v, is defined as the distance traveled, Δs, divided by the elapsed time, Δt;

$$v = \frac{\Delta s}{\Delta t} \qquad (13\text{-}1)$$

A body that travels equal distances in equal times has constant velocity. The average velocity of such a body is equal to the constant velocity.

The surface of an air track has many small holes with air blowing through them. A glider placed on an air track is lifted about a millimeter off the surface by the air; it literally rides on air and moves with virtually no friction.

A spark timer is a devise that generates electrical pulses at a set time interval. It is used together with a special tape, spark timer tape, that has a coating on it that records the location of the sparks that pass through it. The number of sparks per unit time generated by the spark timer is determined by its frequency setting. A frequency of 10 Hertz (Hz) means that there are 10 sparks generated per second so that the time interval between sparks is $1/10^{th}$ second.

In this experiment a glider will be set in motion on a level air track that is wired to a spark timer. The position of the glider will be recorded on spark timer tape at a set repetitive interval. The position of the glider will be plotted as a function of the time. The slope of this line will then be determined. The slope is the velocity.

Procedure

CAUTION - HIGH VOLTAGE

The spark timer is to be left unplugged until you are instructed to plug it in.

1. Position a piece of spark timer tape along the length of the air track in the allotted slot. Use masking or scotch tape to secure the ends of the spark timer tape.

2. Secure a rubber band between a pair of notches on the launcher.

3. Turn on the blower to the air track.

4. Place a glider on the air track.

5. The horizontal wires along the top of the air track are spark wires. Adjust the top of the wire on the side of the glider so that it is few millimeters from the spark wire connected to the spark timer. Adjust the bottom of the wire on the glider so that it is a few millimeters above the spark tape. Set the unplugged spark timer to 10 Hz.

6. Pull the glider against the rubber band on the launcher by means of the cord tied to the top of the glider.

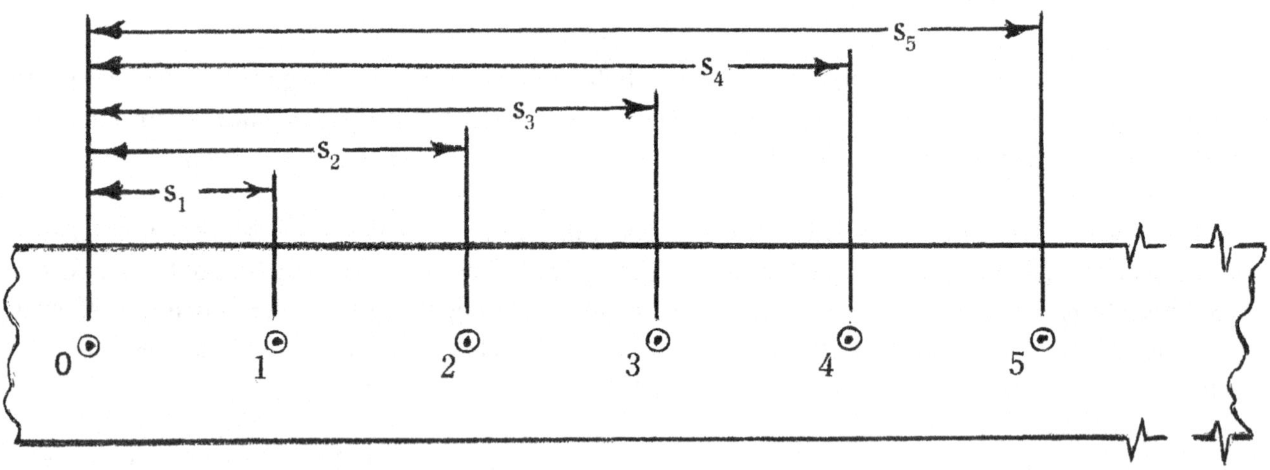

Figure 13-1 Measuring The Distances On The Spark Timer Tape

CAUTION - HIGH VOLTAGE

The spark timer is turned on in the next step. Do not touch any of the wires connected to the spark timer. Do not touch any of the spark wires on the air track.

7. Plug in the spark timer and turn it on. Depress the spark switch. Release the glider. Just before the glider reaches the end of the air track release the spark switch. Turn off the spark timer and unplug it.

8. Remove the spark tape from the air track and examine it. The dots should all be evenly spaced. If any dots are missing the tape is not usable and you need to repeat the experiment up to this point to obtain a usable spark tape.

9. Circle every dot. Label the first circled dot "0" and all the following circled dots "1", "2", "3", ... up to "10". See Figure 13 -1.

10. Measure the distance from dot "0" to each of the other numbered dots as shown in Figure 13-1. Record the distance from dot "0" to dot "1" as s_1, the distance from dot "0" to dot "2" as s_2, the distance from dot "0" to dot "3" as s_3, etc. Record your results in Table 13-1

11. The time interval between the sparks generated by the spark timer was 1/10th second. The elapsed time to each successive circled dot increased by 0.10 second. Therefore, the elapsed time to dot s_1 was 0.1 s; the elapsed time to dot s_2 was 0.2 s; etc. Calculate and record the elapsed times for all the distances s_1 to s_{10}.

12. Plot the elapsed time on the x-axis and the distance traveled on the y-axis.

13. Draw your best straight line for your data. Determine the slope of your line; this is the velocity of the glider. Record your velocity in Table 13-1

Questions

1. How would your results differ if a thicker rubber band were used to launch your glider?

2. If your air track were not perfectly level, what effect would it have on the motion of the glider?

EXPERIMENT 14

KINEMATICS - CONSTANT ACCELERATION

Objective

To study motion with constant acceleration

Equipment

Air track, glider with an 8 inch long cord secured at its top, blower, spark timer, masking or scotch tape, rubber band, small block (about 1.0 inch high)

Discussion and Theory

The description of motion is called kinematics. In this experiment we will study those quantities that are needed to describe motion. In the prior experiment, we defined the *average velocity*, v_{ave}, as

$$v_{ave} = \frac{\text{Distance traveled}}{\text{Elapsed Time}} = \frac{\Delta s}{\Delta t} . \qquad (14\text{-}1)$$

The *instantaneous velocity*, v, is obtained from equation (14-1) for very small values of Δt. (elapsed time approaches zero). In this experiment, whenever we speak of the velocity it is understood that we will mean the instantaneous velocity.

The *average acceleration*, a_{ave}, is defined as

$$a_{ave} = \frac{\text{Change in Velocity}}{\text{Elapsed Time}} = \frac{\Delta v}{\Delta t} . \qquad (14\text{-}2)$$

A body whose velocity changes equal amounts in equal times has a *constant acceleration* and average acceleration equals the constant acceleration. In this experiment we will limit ourselves to constant acceleration.

The quantities needed to study motion with constant acceleration are the distance traveled (s), the final velocity (v_f), the acceleration (a), the elapsed time (t), and the velocity at time t = 0 called the initial velocity (v_i). The relationship between these quantities may be written with four equations;

$$v_f^2 = v_i^2 + 2as \quad (14\text{-}3)$$

$$v_f = v_i + at \quad (14\text{-}4)$$

$$s = (1/2)(v_i + v_f)\,t \quad (14\text{-}5)$$

$$s = v_i t + (1/2)at^2 \quad (14\text{-}6)$$

In this experiment we will tilt an air track slightly so that a glider will undergo an acceleration as it travels down an incline. By means of a spark timer we will record the position of the glider as a function of time. From an analysis of this data we will determine the acceleration of the glider and its velocity at various positions. We will then verify that the above four kinematic equations apply to the motion of the glider.

Procedure

CAUTION - HIGH VOLTAGE

The spark timer is to be left unplugged until you are instructed to plug it in.

1. Measure and record in the bottom of Table 14-1 the distance, L, from the front leg to the back legs of the air track.

2. A small block will be used to lift the single leg. Measure and record in the bottom of Table 14-1 the height, H, of the block in centimeters. Place this block under the single leg of the air track so that the single leg is lifted a height H above the laboratory table.

2. Position a piece of spark timer tape along the length of the air track in the allotted slot. Use masking or scotch tape to secure the ends of the spark timer tape.

4. Turn on the blower to the air track.

5. Place a glider on the air track.

6. The horizontal wires along the side of the air track are spark wires. Adjust the wire on the side of the glider so that it is a few millimeters from the spark wire connected to the spark timer. Adjust the bottom of the wire on the glider so that it is a few millimeters above the spark tape. Set the unplugged spark timer to 10 Hz.

7. Use the cord to hold the glider near the top of the incline.

CAUTION - HIGH VOLTAGE

The spark timer is turned on in the next step. Do not touch any of the wires connected to the spark timer. Do not touch any of the spark wires on the air track.

8. Plug in the spark timer and turn it on. Depress the spark switch. Release the glider. Just before the glider reaches the end of the air track release the spark switch. Turn off the spark timer and unplug it.

9. Remove the spark tape from the air track and examine it. The distance between the dots should increase as the glider traveled down the incline. If any dots are missing the tape may not be usable and you need to repeat the experiment up to this point to obtain a usable spark tape.

10. Choose one dot near the side of the tape corresponding to the top of the incline. Circle and label this first dot "0". Starting with dot "0", circle every dot in the direction down the incline and label these circled dots "1", "2", "3", ... up to "23". See Figure 14-1.

11. Measure the distance from dot "0" to each of the other numbered dots as shown in Figure 14-1. Record the distance from dot "0" to dot "1" as s_1, the distance from dot "0" to dot "2" as s_2, the distance from dot "0" to dot "3" as s_3, etc. Record your results in Table 14-1.

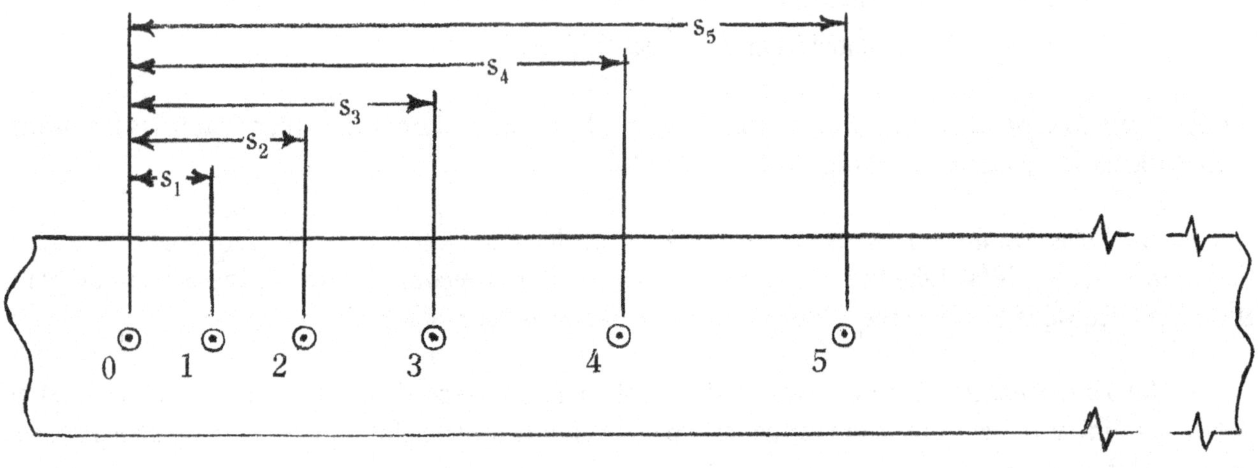

Figure 14-1 Measuring the Distances on the Spark Timer Tape

12. Calculate and record in Table 14-1 the distances Δs between neighboring odd dots by subtracting the distances found in step 11 in the following way: $\Delta s_2 = s_3 - s_1$, $\Delta s_4 = s_5 - s_3$, $\Delta s_6 = s_7 - s_5$, etc. The distances Δs are indicated in Table 14-1 by even numbers because these Δs's will be used to determine the velocity of the glider at the location of the even numbered dots.

13. Calculate the average velocity between neighboring odd numbered dots by dividing the Δs's by the time interval between the odd dots (0.2 second). Record these as the V's in Table 14-1. These average velocities are the actual velocities of the glider at the even numbered dots (i.e., v_2 is the velocity at dot #2 with t = 0.2 s, v_4 is the velocity at dot #4 with t = 0.4 s, etc.)

14. Plot a graph of the velocities found in step 13 (i.e., the V's in Table 14-1) as a function of the elapsed time. Draw the best straight line. Calculate the slope of this line. This is your experimental value for the acceleration of the glider. Record this value at the bottom of Table 14-1.

15. For the small incline used in this experiment, the theoretical acceleration of the glider is calculated from

$$a = g\,(H/L)$$

where g is 980 cm/s^2, H is the height of the block measured in step 2, and L is the distance from the front leg to the back legs measured in step 1. Calculate the theoretical acceleration and the percent error in your experimental acceleration. Record your results in Table 14-1.

16. Verify the kinematic equations as follows: (Record all values in Table 14-2)

(a) Let

a = the calculated acceleration from your graph of V vs T.

$v_i = V_2$ (from Table 14-1)

$v_f = V_{22}$ (from Table 14-1)

$t = T_{22} - T_2$ (calculated from the values in Table 14-1)

$s = s_{22} - s_2$ (calculated from the values in Table 14-1).

We are going to verify that each of these quantities can be calculated from the other quantities by means of the kinematic equations - equations (14-3) to (14-6).

(b) *To calculate v_f from equation (14-3):* Calculate and record $v_i^2 + 2as$ from the values determined in step 16(a) Take the square root of this result and record. Determine the percent error in this calculated value of v_f. Use the value of v_f from step 16(a) as your standard.

(c) *To calculate v_f from equation (14-4):* Calculate and record $v_i + at$ from the values in step 16(a). Determine the percent error in this calculated value of v_f. Use the value of v_f from step 16(a) as your standard.

(d) *To calculate s from equation (14-5):* Calculate and record $(1/2)(v_i + v_f)t$ from the values in step 16(a). Determine the percent error in this calculated value of s. Use the value of s from step 16(a) as your standard.

(e) *To calculate s from equation (14-6):* Calculate and record $v_i t + (1/2)at^2$ from the values in step 16(a). Determine the percent error in this calculated value of s. Use the value of s from step 16(a) as your standard.

Questions

1. A car stars from rest ($v_i = 0$) and accelerates 3.0 m/s^2 for a distance of 30 m. What is its final velocity?

2. A train accelerates 2.0 m/s^2 from an initial velocity of 10 m/s for 8 seconds. What is its final velocity.

3. How far will a car travel if it accelerates uniformly from 10 m/s to 30 m/s in 20 seconds?

4. A body is dropped ($v_i = 0$). It accelerates 9.8 m/s^2 for 3.0 seconds. What is its final velocity?

Name____________________ Date Performed____________________

Lab. Partners ____________________ Date Submitted____________________

____

____

Table 14-1
Determination of the Velocities of the Glider

Dot No.	Elapsed Time (s)	Distance Fallen (cm)	Δs's(cm)	Velocity(cm/s)
0	0.0			
1	0.1	s_1 = ________		
2	0.2	s_2 = ________	$\Delta s_2 = s_3 - s_1$ = ________	$V_2 = (\Delta s_2 / \Delta t)$ = ________
3	0.3	s_3 = ________		
4	0.4	s_4 = ________	$\Delta s_4 = s_5 - s_3$ = ________	$V_4 = (\Delta s_4 / \Delta t)$ = ________
5	0.5	s_5 = ________		
6	0.6	s_6 = ________	$\Delta s_6 = s_7 - s_5$ = ________	$V_6 = (\Delta s_6 / \Delta t)$ = ________
7	0.7	s_7 = ________		
8	0.8	s_8 = ________	$\Delta s_8 = s_9 - s_7$ = ________	$V_8 = (\Delta s_8 / \Delta t)$ = ________
9	0.9	s_9 = ________		
10	1.0	s_{10} = ________	$\Delta s_{10} = s_{11} - s_9$ = ________	$V_{10} = (\Delta s_{10} / \Delta t)$ = ________
11	1.1	s_{11} = ________		
12	1.2	s_{12} = ________	$\Delta s_{12} = s_{13} - s_{11}$ = ________	$V_{12} = (\Delta s_{12} / \Delta t)$ = ________
13	1.3	s_{13} = ________		
14	1.4	s_{14} = ________	$\Delta s_{14} = s_{15} - s_{13}$ = ________	$V_{14} = (\Delta s_{14} / \Delta t)$ = ________
15	1.5	s_{15} = ________		
16	1.6	s_{16} = ________	$\Delta s_{16} = s_{17} - s_{15}$ = ________	$V_{16} = (\Delta s_{16} / \Delta t)$ = ________
17	1.7	s_{17} = ________		
18	1.8	s_{18} = ________	$\Delta s_{18} = s_{19} - s_{17}$ = ________	$V_{18} = (\Delta s_{18} / \Delta t)$ = ________
19	1.9	s_{19} = ________		
20	2.0	s_{20} = ________	$\Delta s_{20} = s_{21} - s_{19}$ = ________	$V_{20} = (\Delta s_{20} / \Delta t)$ = ________
21	2.1	s_{21} = ________		
22	2.2	s_{22} = ________	$\Delta s_{22} = s_{23} - s_{21}$ = ________	$V_{22} = (\Delta s_{22} / \Delta t)$ = ________
23	2.3	s_{23} = ________		

Distance from front to back legs of air track: L = ____________ Height of block: H = ____________

Experimental Acceleration = ________ Theoretical Acceleration = ________ % error =________

Name________________________________

Table 14-2
Verification of the Kinematic Equations

a = ______________________________

v_i = ______________________________

v_f = ______________________________

t = ______________________________

s = ______________________________

Verification of equation (14-3); $v_f^2 = v_i^2 + 2as$

$v_i^2 + 2as$ = ____________ $\sqrt{v_i^2 + 2as}$ = ____________ % error = ____________

Verification of equation (14-4); $v_f = v_i + at$

$v_i + at$ = ____________ % error = ____________

Verification of equation (14-5); $s = (1/2)(v_i + v_f)\,t$

$(1/2)(v_i + v_f)\,t$ = ____________ % error = ____________

Verification of equation (14-6); $s = v_i t + (1/2)at^2$

$v_i t + (1/2)at^2$ = ____________ % error = ____________

EXPERIMENT 15

THE MECHANICAL ADVANTAGE OF THE INCLINED PLANE

Objective

To investigate the mechanical advantage of an inclined plane.

Equipment

Cart, pulley, cord, slotted weights, inclined plane (a flat surface elevated at one end), mass hanger, balance

Discussion and Theory

In Figure 15-1, we see an inclined plane of length L and height H. If a net force, F, is needed to keep the weight, W, from just sliding down the incline, then *the mechanical advantage*, M.A., is defined as:

$$M.A. = \frac{W}{F} . \qquad (15\text{-}1)$$

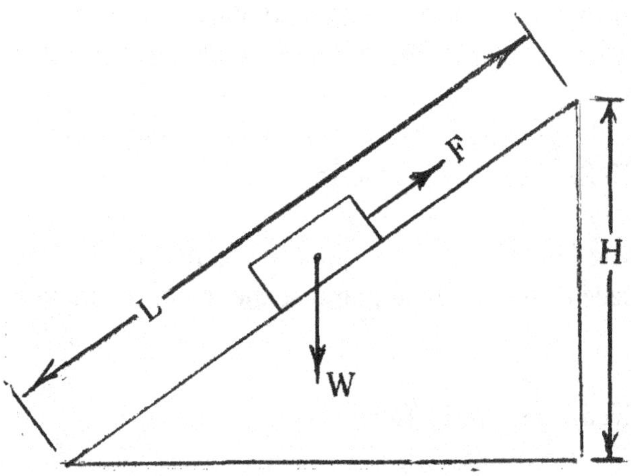

Figure 15-1 The Mechanical Advantage of an Inclined Plane

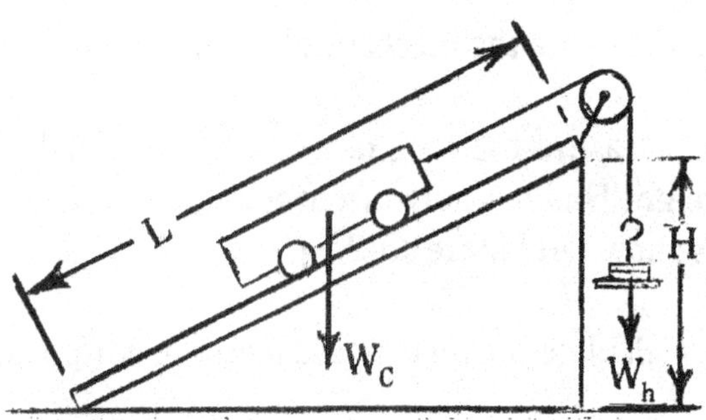

Figure 15-2 Experimental Set-Up to Determine the Mechanical Advantage

In this experiment, we will use a cart placed on an incline and held stationary by means of a hanging weight, W_h, as shown in Figure 15-2. If the weight of the cart is W_C, we get that:

$$M.A. = \frac{W_C}{W_h} \qquad (15\text{-}2)$$

The mechanical advantage of an inclined plane can be shown to be equal to the ratio of the length of the inclined plane, L, to the height of the inclined plane, H (See Figures 15-1 and 15-2). Therefore, we may write:

$$M.A. = \frac{L}{H} \qquad (15\text{-}3)$$

We will verify equation (15-3) in this experiment.

Procedure

1. Use a balance to determine the mass of the cart in kilograms. Calculate the weight in Newtons of the cart by multiplying the mass by 9.8 m/s^2. Record your values in Table 15-1.

2. Place the cart on the inclined plane. Attach a cord to the cart as shown in Figure 15-2. Pass the cord over the pulley. The cord should be parallel to the inclined surface. Attach a mass hanger to the free end of the cord. On this mass hanger add just enough mass to hold the cart stationary. Calculate the weight of the total hanging mass (including the hanger). Record the amount of hanging mass in kilograms and its weight in Newtons in Table 15-1.

3. Calculate the mechanical advantage from equation (15-2). Record in Table 15-1.

4. Measure the length, L, of the inclined plane and the height, H, of the inclined plane. (Note The height H is measured from the table to the bottom surface of the inclined plane at the top of the incline as shown in Figure 15-2.)

5. Calculate the mechanical advantage from equation (15-3). Record in Table 15-1

6. Compare the mechanical advantage calculated from equation (15-2) to the mechanical advantage calculated from equation (15-3) by finding the percent error. Use the result from equation (15-2) as the

standard value. Record you result in Table 15-1.

7. Repeat steps 1 through 6 for as many runs as your instructor may indicate. Your instructor may ask you to change the angle of the incline and/or to add mass to the cart.

Questions

1. Why is it important that the cord be parallel to the inclined surface as it pulls on the cart?

2. Would the mechanical advantage be affected if a heavier cart were used in this experiment?

3. When you measured the height H of the incline, why were you told to measure to the bottom surface of the inclined plane for the height of the incline, rather then to the top surface of the inclined plane?

Name______________________________ Date Performed______________________
Lab. Partners ______________________ Date Submitted______________________
____
____

Table 15-1
Calculation of the Mechanical Advantage of Incline Planes

	First Incline	Second Incline	Third Incline
Mass of the cart (M_C)	________	________	________
Weight of Cart (W_C)	________	________	________
Total Hanging Mass (M_h)*	________	________	________
Total Hanging Weight (W_h)	________	________	________
Mechanical Advantage (W_C / W_h)	________	________	________
Length of Inclined Plane (L)	________	________	________
Height (H)	________	________	________
Mechanical Advantage (L/H)	________	________	________
Percent Error	________	________	________

* Include the mass of the platform

EXPERIMENT 16

THE MECHANICAL ADVANTAGE OF LEVERS

Objective

To investigate the mechanical advantage of levers and the condition for rotational equilibrium.

Equipment

A meter stick, two slip-on mass hangers, one knife-edge support with stand, one set of hooked masses.

Discussion and Theory

In Figure 16-1(a) a force F is applied perpendicular to a rod a distance L from a pivot point P. *The tendency of a force to cause rotation about a point is called the torque.* If the distance L is perpendicular to the force F it is called the *moment arm. To calculate the torque, the force F is multiplied by the moment arm L*

$$\text{Torque} = F \times L. \tag{16-1}$$

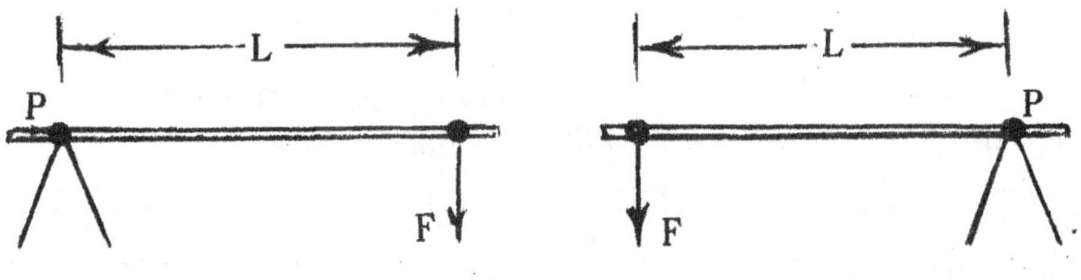

(a) A clockwise torque (b) A counterclockwise torque

Figure 16-1 Torques

The torque due to the force F in Figure 16-1(a) has the tendency to cause a clockwise rotation about the point P and is referred to as a *clockwise torque.* The torque due to the force F in Figure 16-1(b) has the tendency to cause a *counterclockwise* rotation about the point P and is referred to as a *counterclockwise torque.*

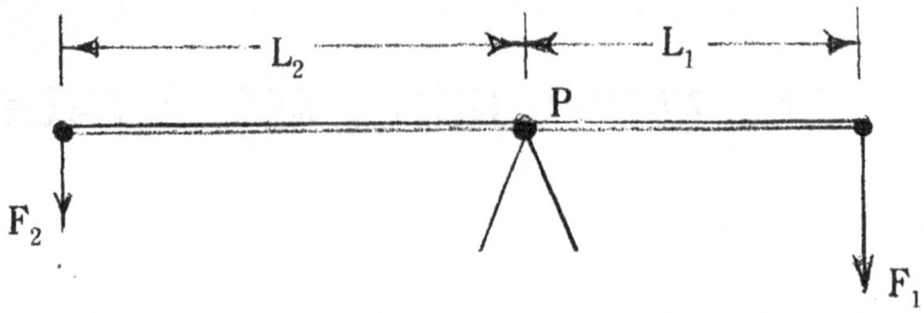

Figure 16-2 Two Torques in Rotational Equilibrium

In Figure 16-2 there are two forces acting on the rod. Force F_1 exerts a clockwise torque, $F_1 \times L_1$, about the point P, while force F_2 exerts a counterclockwise torque, $F_2 \times L_2$, about point P. *When the sum of all clockwise torques equals the sum of all counterclockwise torques, then the object is said to be in rotational equilibrium.* A stationary object in rotational equilibrium will not start to rotate even though there are forces acting on it. For the object in Figure 16-2 to be in rotational equilibrium

$$F_1 \times L_1 = F_2 \times L_2 \tag{16-2}$$

When a force F_2 counterbalances the effect of another force F_1 to cause rotation there is a mechanical advantage,

$$\text{M.A.} = \frac{F_1}{F_2}. \tag{16-3}$$

The system shown in Figure 16-2 is a *lever*. Let us assume that the torque due to force F_1 is counterbalanced by the torque due to force F_2. From equation (16-2), we get:

$$\frac{F_1}{F_2} = \frac{L_2}{L_1} \tag{16-4}$$

Here F_1 / F_2, is the mechanical advantage and therefore L_2 / L_1 is also the mechanical advantage. In this experiment we will verify equation (16-4).

Procedure

Part I

1. Position the center of a meter stick through the knife-edge support so that the thumbscrew of the support is along the bottom of the meter stick. Place the knife-edge on the support bracket at pivot point P as shown in Figure 16-3. If the meter stick does not balance, reposition the meter stick on the knife-edge support so that it does balance. Tighten the thumbscrew.

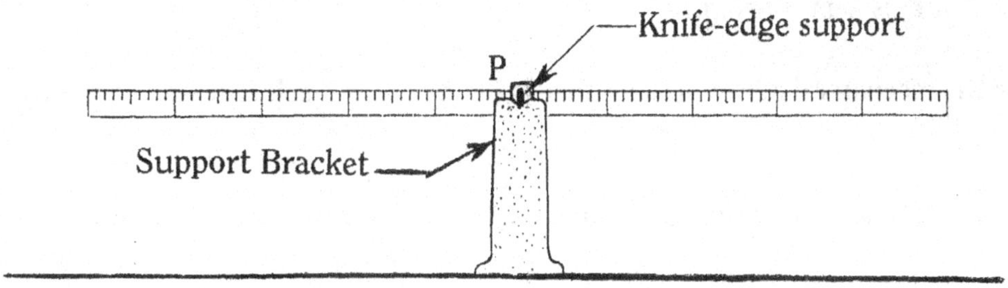

Figure 16-3 Balancing the Meter Stick

2. Determine the mass of the hangers h_1 and h_2 in kilograms and record them as m_1 and m_2 in Table 16-1 (Recall 1 gram = 0.001 kilogram).

3. Place hanger h_1 at L_1 = 0.100 meters (10 centimeters) from point P and tighten its thumbscrew (See Figure 16-4). Suspend 0.050 kilograms (50 g) from h_1. Calculate and record in Table 16-1 the total suspended mass, M_{T1}, by adding the mass of hanger h_1 to the 0.050 kg. The weight of M_{T1} is the force F_1. Calculate the weight of M_{T1} from M_{T1} X 9.8 m/s^2 and record as F_1.

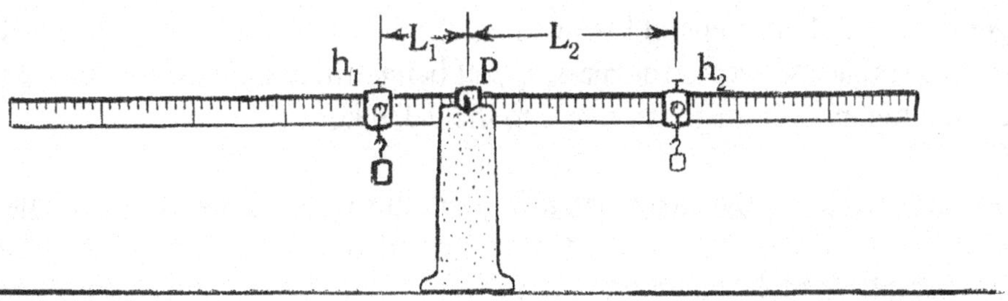

Figure 16-4 The Experimental Set-Up for the Mechanical Advantage of Levers

4. Place hanger h_2 on the other end of the meter stick and suspend 0.050 kg (50 g) from it as shown in Figure16-4. Calculate and record the total suspended mass, M_{T2}, by adding the mass of hanger h_2 to the 0.050 kg. The weight of M_{T2} is the force F_2. Calculate the weight of M_{T2} from M_{T2} X 9.8 m/s^2 and record as F_2.

5. Slide hanger h_2 along the meter stick until the system is balanced. Record the distance from P to hanger h_2 as L_2 (See Figure 16-4).

6. Keep h_1 at 10.0 centimeters from point P and repeat steps 3, 4 and 5 five more times with larger masses suspended from h_1; first with 0.100 kg suspended from h_1, then with 0.150 kg, then with 0.200 kg, then with 0.250 kg and finally with 0.300 kg suspended. Keep 0.050 kg suspended from h_2 during each repetition so that F_2 is kept constant.

7. Calculate the mechanical advantages by dividing F_1 by F_2 for each case and record the values in Table 16-1.

8. Calculate the mechanical advantages by dividing L_2 by L_1 for each case and record the values in Table 16-1.

9. Assume F_1 / F_2 is the standard value. Determine the percent error in the mechanical advantages calculated from L_2 /L_1 and record in Table 16-1.

Part II

1. The point at which the meter stick balances is its center of mass. Find the position of the center of mass of the meter stick and record in Table 16-2.

2. Move the meter stick through the knife-edge support so that the knife-edge supports it 0.100 m (10 cm) to one side of the center of mass (c.m.) as shown in Figure 16-5.

3. Place hanger h_2 on the shorter end of the meter stick with 0.050 kg (50 g) suspended from it as shown in Figure 16-5. Determine and record the mass, m_2, of hanger h_2 .Calculate and record the total hanging mass, M_2, by adding the mass of the hanger, m_2 , to the 0.050 kg.

4. Locate the hanger, h_2, on the meter stick so that the meter stick is again balanced. Record the distance L_2 from the pivot point to the hanger h_2. Calculate and record the total force F_2 due to the weight of M_2 by multiplying M_2 by 9.8 m/s^2 .

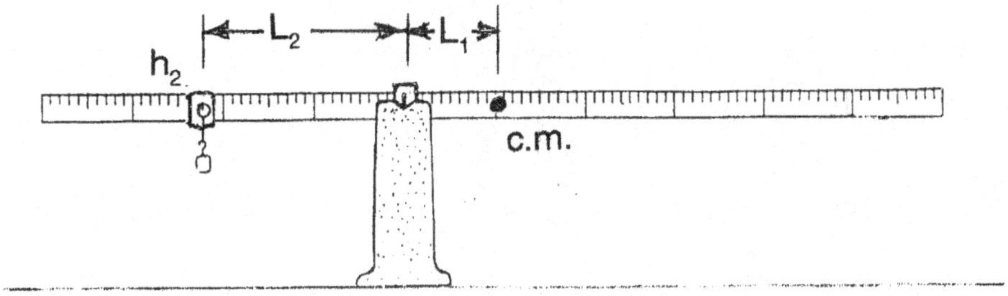

Figure 16-5 Finding the Weight of the Meter Stick from Torques

5. If we let the weight, W_S, of the meter stick be the force F_1 and the moment arm to the center of mass (which is 0.100 m) be L_1 in equation (16-4), then

$$W_S = \frac{L_2}{L_1} F_2. \qquad (16\text{-}5)$$

Using this relationship, calculate the weight of the meter stick.

6. Determine the actual mass, M_S', of the meter stick from a balance and record in Table 16-2.

7. Calculate the actual weight, W_S', of the meter sick by multiplying M_S' by 9.8 m/s^2 and record in Table 16-2.

8. Determine the percent error of the calculated weight in step 4 and record. The actual weight is the standard value.

Question

1. Using the concept of the mechanical advantage, explain how a 50 pound boy can lift a 150 pound man on a see-saw.

Name ______________________ Date Performed ______________________

Lab. Partners ______________________ Date Submitted ______________________

____

____

Table 16-1
Finding the Mechanical Advantages

m_1 = ______________ m_2 = ______________

M_{T1} (kg)	F_1 (N)	L_1 (m)	M_{T2} (kg)	F_2 (N)	L_2 (m)	F_1 / F_2	L_2 / L_1	% Error
		0.100						
		0.100						
		0.100						
		0.100						
		0.100						

Table 16-2
Determining the Weight of the Meter Stick from Torques

Location of c.m. of meter stick = ______________

L_I = ______________ 0.100 m ______________

m_2 = ______________

M_2 = ______________

F_2 = ______________

L_2 = ______________

W_s (Calculated weight of meter stick) = ______________

Actual mass of the meter stick, M_s' = ______________

Actual weight of meter stick, W_s' = ______________

Percent Error = ______________

EXPERIMENT 17

THE MECHANICAL ADVANTAGE OF PULLEYS

Objective

To investigate the mechanical advantage of pulleys.

Equipment

Cord, slotted weights, mass platforms, several pulleys.

Discussion and Theory

A system of pulleys can be used to lift heavy objects. The *mechanical advantage of a system of pulleys is equal to the number of ropes that support the movable pulley.* Figure 17-1 shows several pulley systems and their respective mechanical advantages. The weight, W_1, on platform I, is supported by the pulley system. The weight, W_2, on platform II, supports the total weight on platform I plus the weight of the movable pulleys.

The mechanical advantage of each pulley system in this experiment will be found from the slope of the line obtained by plotting W_1 as a function of W_2. The weight of the movable pulleys only has the effect of introducing a constant to the equation of the line and does not affect the slope.

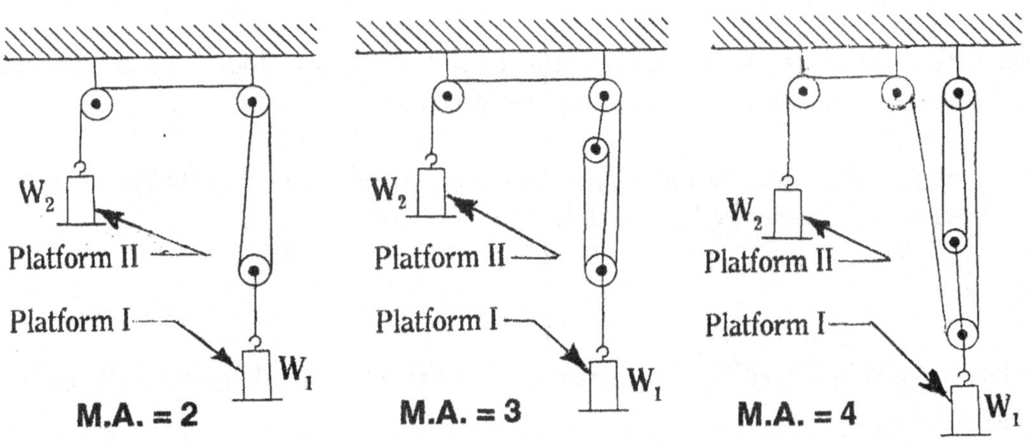

Figure 17-1 The Mechanical Advantages of Several Pulley Systems

Procedure

1. There are several set-ups of pulleys with various mechanical advantages. Choose a set-up.

2. Determine the mechanical advantage of your set-up from the number of ropes supporting the movable pulley(s).

3. As shown in Figure 17-1 the platform attached to the movable pulley system is referred to as platform I; the platform attached to the fixed pulley is referred to as platform II. Place 0.250 kg (250 g) on platform I of the pulley set-up. The total mass M_1 is 0.300 kg (0.250 kg plus 0.050 kg for the platform).

4. Add slotted masses to platform II until the system is balanced. The total mass M_2 is the mass you have added to the platform plus the mass of the platform which is 0.050 kg. Record the total mass M_2 in Table 17-1.

5. Increase the total mass M_1 in steps of 0.100 kg from 0.300 kg to 0.900 kg. In each case add the proper mass to platform II to balance the system. For each case record the total mass M_2 of platform II (mass of platform plus slotted masses) in Table 17-1.

6. Calculate the weights W_1 of all the masses M_1 and the weights W_2 of all the masses M_2 by multiplying each mass by 9.8 m/s^2. Record your results in Table 17-1.

7. Repeat steps 2 through 6 for as many set-ups as your instructor requires. Use Tables 17-2 through 17-4, as required.

8. Determine your experimental mechanical advantage for each pulley system as follows:

(a) Plot the weights W_1 on the y-axis and the weights W_2 on the x-axis.

(b) Determine the slope of the best straight line of these points. This is your experimental mechanical advantage. Record you result in the appropriate table.

(c) Compare this with the mechanical advantage found from the number of ropes. Use the number of ropes as your standard value. Find the percent error.

Questions

1. When the system is "balanced" it is found that the addition or removal of several grams has no effect. Why is this so?

2. Theoretically, the point where your line crosses the y-axis is the weight of your movable pulley(s). Does this value of your y-intercept seem reasonable for the weight of the movable pulley(s)?

Name________________________________ Date Performed________________________
Lab. Partners ________________________ Date Submitted________________________
____
____

Table 17-1
Mechanical Advantage of the First Pulley System

Mechanical Advantage (From the number of ropes):______________

Platform I		Platform II	
Mass M_1 (kg)	Weight W_1 (N)	Mass M_2 (kg)	Weight W_2 (N)
0.300			
0.400			
0.500			
0.600			
0.700			
0.800			
0.900			

Mechanical Advantage (From slope) = ______________
Percent Error in slope =________________________

Table 17-2
Mechanical Advantage of the Second Pulley System

Mechanical Advantage (From the number of ropes):______________

Platform I		Platform II	
Mass M_1 (kg)	Weight W_1 (N)	Mass M_2 (kg)	Weight W_2 (N)
0.300			
0.400			
0.500			
0.600			
0.700			
0.800			
0.900			

Mechanical Advantage (From slope) = ______________
Percent Error in slope =________________________

Name________________________________

Table 17-3
Mechanical Advantage of the Third Pulley System

Mechanical Advantage (From the number of ropes):______________

Platform I		Platform II	
Mass M_1 (kg)	Weight W_1 (N)	Mass M_2 (kg)	Weight W_2 (N)
0.300	______	______	______
0.400	______	______	______
0.500	______	______	______
0.600	______	______	______
0.700	______	______	______
0.800	______	______	______
0.900	______	______	______

Mechanical Advantage (From slope) = ______________
Percent Error in slope =________________________

Table 17-4
Mechanical Advantage of the Fourth Pulley System

Mechanical Advantage (From the number of ropes):______________

Platform I		Platform II	
Mass M_1 (kg)	Weight W_1 (N)	Mass M_2 (kg)	Weight W_2 (N)
0.300	______	______	______
0.400	______	______	______
0.500	______	______	______
0.600	______	______	______
0.700	______	______	______
0.800	______	______	______
0.900	______	______	______

Mechanical Advantage (From slope) = ______________
Percent Error in slope =________________________

EXPERIMENT 18

THE ACCELERATION DUE TO GRAVITY

Objective

To determine the acceleration due to gravity.

Equipment

The Behr Free Fall Apparatus, a spark timer, a power supply, a cut off switch, wires, a meter stick.

Discussion and Theory

The acceleration, a, is defined as the change in velocity per unit time. Therefore,

$$a = \frac{\Delta v}{\Delta t} \tag{18-1}$$

where Δv is the change in velocity and Δt is the elapsed time.

Whenever a body falls freely on the surface of the earth, its acceleration, called the acceleration due to gravity, g, is found to be:

$$\begin{aligned} g &= 9.8\ \text{m/s}^2 \\ &= 980\ \text{cm/s}^2 \\ &= 32\ \text{ft/s}^2. \end{aligned}$$

In this experiment we will make use of the Behr Free Fall Apparatus shown in Figure 18-1. An electromagnet is used to suspend a mass. The electromagnet is connected to a power supply with a cut off switch in its circuit. The suspended mass has a conductive ring around it. When the cut off switch is opened, power ceases to go to the electromagnet and the mass starts to fall freely. The freely falling mass passes between two wires connected to a spark timer. A strip of spark timer tape is located between the freely falling mass and one of the spark wires. Each $1/60^{th}$ second the spark timer generates a spark which jumps from one spark wire to the conducting metal ring on the freely falling mass, through the spark timer tape and to the other spark wire. The spark paper records each spark as a spot, which gives the location of the freely falling mass at intervals of $1/60^{th}$ second.

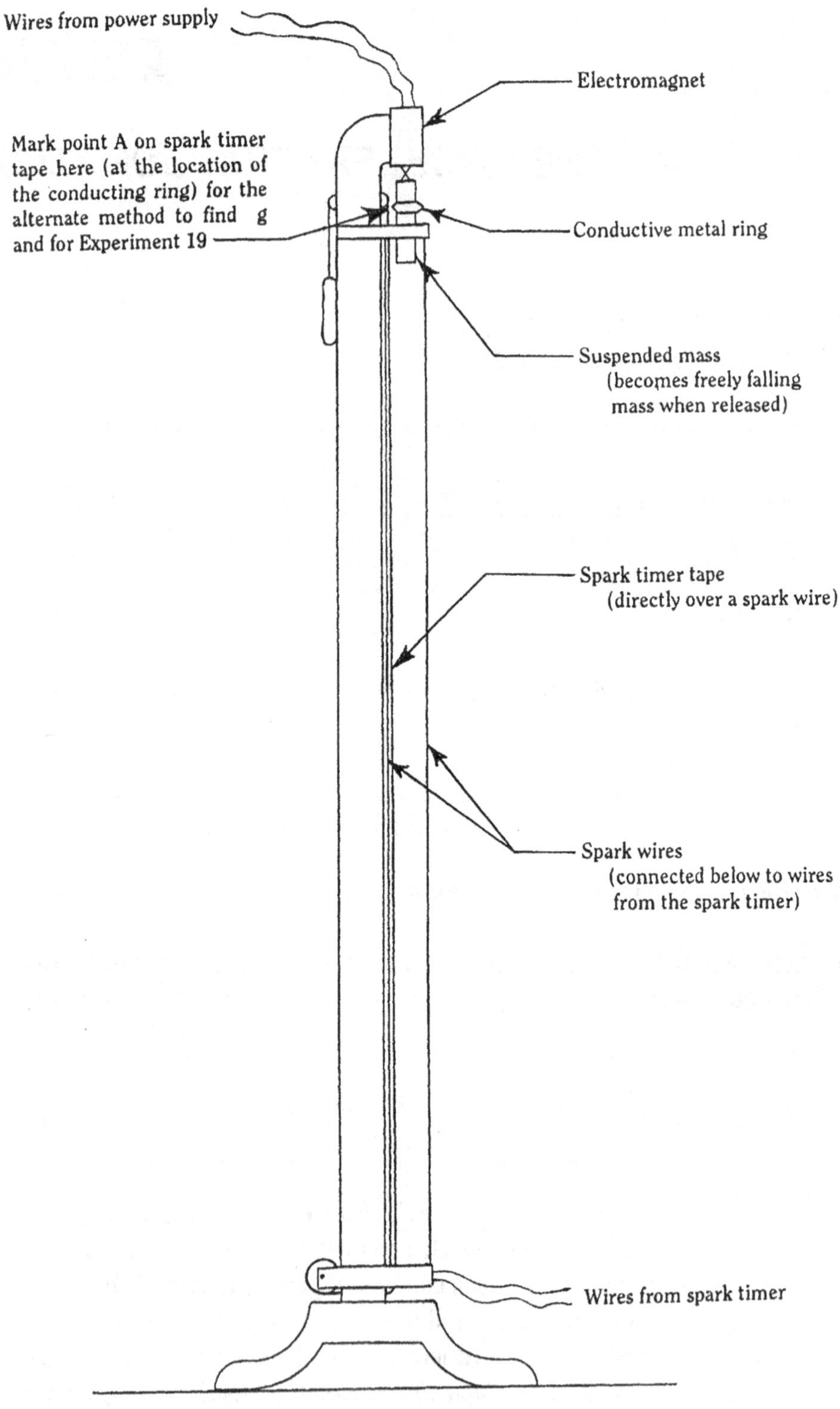

Figure 18-1 The Behr Free Fall Apparatus

The spark timer tape is analyzed to determine the distance traveled as a function of the elapsed time. From this data the velocity as a function of time is determined and plotted on a graph. The acceleration due to gravity is then determined from the slope of the line.

Alternate Method of Determining the Acceleration Due to Gravity

The distance, s, that a freely falling body falls in a time, t, is

$$s = v_0 t + (1/2) gt^2 \qquad (18\text{-}2)$$

where v_0 is the velocity at time t = 0. If the distance is measured from where the freely falling body is released at t = 0, then $v_0 = 0$ and

$$s = (1/2) g t^2 \qquad (18.3)$$

The square root of this equation is

$$s^{1/2} = (g/2)^{1/2} t \qquad (18\text{-}4)$$

If $s^{1/2}$ is plotted as a function of the time, the slope, m, of the line should be

$$m = (g/2)^{1/2} . \qquad (18\text{-}5)$$

Procedure

1. The free fall apparatus should be properly leveled and wired prior to the start of the experiment. A power supply should be wired in series with a switch and the electromagnet of the Behr Free Fall Apparatus. The spark timer should not yet be plugged in and should be connected to the spark wires near the bottom of the apparatus. The spark timer switch should be open and the spark timer set at 60 Hz. The apparatus should have spark timer tape threaded through it.

2. Energize the electromagnet at the top of the apparatus and suspend the mass from the electromagnet without swinging back and forth, as shown in Figure 18-1. If the alternate method of determining the acceleration due to gravity is to be used, mark the location of the conducting metal ring of the suspended mass onto the spark timer tape. Label this point "A".

CAUTION - HIGH VOLTAGE

THE SPARK TIMER IS TURNED ON IN STEP 3. A HIGH VOLTAGE IS PRESENT IN THE WIRES CONNECTED TO THE SPARK TIMER WHEN THE SPARK TIMER IS TURNED ON.

3. Plug in the spark timer. Push the "on" switch down for a second to be certain that there is a spark.

4. Having verified that the apparatus functions properly, hold the spark timer switch in the "on" position and open the switch to the electromagnet. This releases the suspended mass so that it falls freely. Release the "on" switch of the spark timer when the freely falling mass hits the bottom.

5. Unplug the spark timer.

6. Remove the strip of spark timer tape with the record of the free fall.

7. Examine the strip of spark timer tape and check that the spark timer made at least 22 successive spots. If not, do the experiment over again up to this point.

8. Start at the end with the dots closest together. If a point "A" was not marked onto the spark timer tape in step 2, label the first dot "A". Label each dot below "A" with successive numbers ("1", "2", "3", etc.). See Figure 18-2.

9. Measure the distances from A to each successive dot in cm. (i.e., measure the distance s_1 from A to dot #1, the distance s_2 from A to dot #2, the distance s_3 from A to dot #3, etc.). See Figure 18-2.

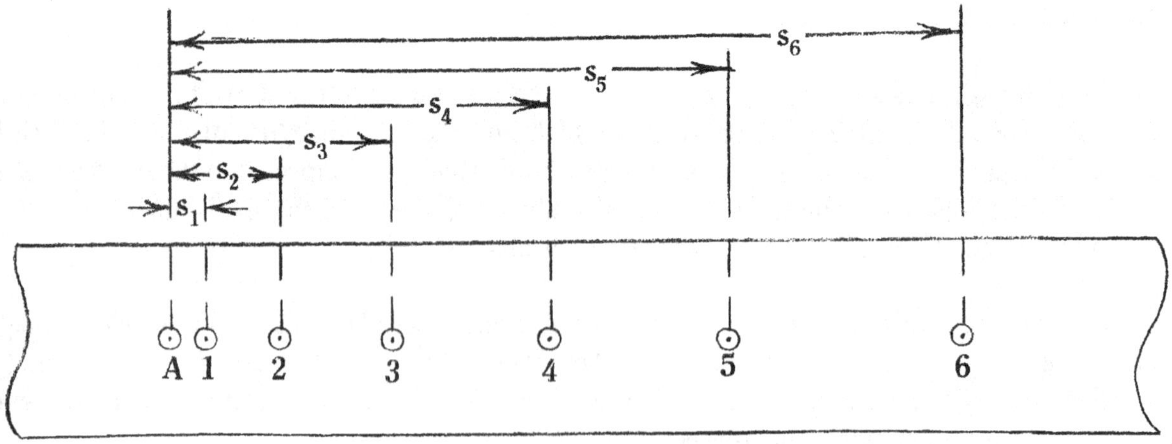

Figure 18-2 Measuring The Distances On The Spark Timer Tape

10. Calculate and record in Table 18-1 the distances Δs between neighboring odd dots by subtracting the distances found in step 9 in the following way: $\Delta s_2 = s_3 - s_1$, $\Delta s_4 = s_5 - s_3$, $\Delta s_6 = s_7 - s_5$, etc. The distances Δs are indicated in Table 18-1 by even numbers because these Δs's will be used to determine the velocity of the freely falling mass at the location of the even numbered dots.

11. Calculate the average velocity between neighboring odd numbered dots by dividing the Δs's by the time interval between the odd dots (1/30th of a second). Record these as the v's in Table 18-1. These average velocities are the actual velocities of the freely falling body at the even numbered dots.

12. Plot a graph of the velocities found in step 11 (i.e., the v's in Table 18-1) as a function of the elapsed time. Draw the best straight line. Calculate the slope of this line. This is your experimental value for the acceleration due to gravity. Record this value at the bottom of Table 18-1, column 5 as g_{exp}. Calculate your percent error. Record in Table 18-1, column 5 under the value of g_{exp} just calculated.

The Following Steps Apply to the Alternate Method of Determining "g"

13. Determine the square root of the distances s_1, s_2, s_3, etc. and enter these values in the last column of Table 18-1.

14. Plot $\sqrt{s}$ on the y-axis and the elapsed time on the x-axis. Draw the best straight line.

15. Find the slope of the line.

16. The acceleration due to gravity is obtained by solving equation (18-5) for g and substituting in the value of the slope found in step 15;

$$g = 2\,m^2 = 2\,(\text{slope})^2\,.$$

Find the percent error. Record your results at the bottom of the last column of Table 18-1 in the spaces provided.

Questions

1. The acceleration, according to equation (18-1), is the change in velocity per unit time. If the time is in seconds it tells us the change in velocity per second. Therefore, if a car accelerates 5 m/s^2 this means that its velocity increases 5 m/s each second. As an object falls by how much does its velocity increase each second?

2. Why did we use only the odd dots to determine the velocities and then plot them at times corresponding to even dots?

EXPERIMENT 19

THE CONSERVATION OF ENERGY

Objective

To examine the conservation of energy of a freely falling body.

Equipment

The Behr Free Fall Apparatus, a meter stick, balance, a switch.

Discussion and Theory

A conservation law tells us that something remains unchanged under certain conditions. *The Principal of the Conservation of Energy says that energy can neither be created nor destroyed; it can only be converted from one form of energy into another form of energy.* Therefore, if we consider a system in which energy neither enters nor leaves, then the total energy of the system must remain constant. In this experiment we will consider the conservation of energy as it applies to a freely falling body. The S.I. unit for energy is the Joule (1 J = 1 N m).

During free fall, two forms of energy are involved, the *potential energy* and the *kinetic energy*. The potential energy depends on position. The kinetic energy depends on velocity. The gravitational potential energy, P.E., of an object is

$$\text{P.E.} = mgh \qquad (19\text{-}1)$$

where m = the mass of the object

g = the acceleration due to gravity
= 9.8 m/s^2

h = the height of the object above some reference point.

The kinetic energy, K.E., of an object is

$$\text{K.E.} = (1/2)\, mv^2 \qquad (19\text{-}2)$$

where m = the mass of the object

v = the speed of the object.

For a freely falling object, the conservation of energy requires that,

$$P.E. + K.E. = \text{constant}, \tag{19-3}$$

or

$$mgh + (1/2)\, mv^2 = \text{constant}. \tag{19-4}$$

The sum of the potential and kinetic energies of a system is called the *mechanical energy*. During free fall the mechanical energy remains constant.

If an object is dropped from a height h_0 and falls to a height h, then the decrease in its potential energy is:

$$\begin{aligned} \text{decrease in P.E.} &= mgh - mgh_0 \\ &= mg(h - h_{0)}) \\ &= mgs \end{aligned} \tag{19-5}$$

where s is the distance fallen, $(h - h_0)$. The conservation of energy requires that the decrease in one form of energy must result in an increase in some other form of energy. In the case of a freely falling body the decrease in the potential energy must equal the increase in the kinetic energy;

$$| mgs | = (1/2)\, mv^2. \tag{19-6}$$

The absolute magnitude of the decrease in the potential energy is written on the left side of the equation since the decrease in potential energy is negative.

In this experiment we will show that as the potential energy decreases the kinetic energy increases so that equation (19-6) is satisfied, and that the mechanical energy of this system remains constant as stated by equation (19-4).

To perform this experiment we will use the Behr Free Fall Apparatus. This is the same apparatus used in performing Experiment 18, *The Acceleration Due To Gravity*. We will suspend a mass from the energized electromagnet of the apparatus as shown in Figure 18-1 of the prior experiment. We will record the location of the ring of the freely falling body on the spark timer tape prior to permitting the mass to fall. A spark timer connected to two spark wires will cause a series of sparks at intervals of $1/60^{th}$ of a second as the mass falls. These sparks will jump from one wire, to the metal ring around the falling body, through the spark timer tape, and finally to the second spark wire. As the sparks pass through the spark timer tape they mark the location of the ring of the falling mass as a series of dots. These dots give the distance fallen, s, by the mass as a function of time.

To analyze the spark timer tape :

(a) We will determine the decrease in the potential energy of the falling mass from equation (19-5).

(b) The velocities at various points along the path of the falling mass will be calculated from the distance traveled between dots divided by the elapsed time. Once these velocities are known the kinetic energies will be determined from equation (19-2).

(c) We will compare the decrease in the potential energy to the kinetic energy, to see if they are equal as required by equation (19-6).

(d) The height above a reference point near the bottom of the spark timer tape will be calculated for those points where the kinetic energy has been determined. From these heights the potential energy of the falling mass will be determined at each height from equation (19-1).

(e) The mechanical energy of the falling mass will be determined at various points by adding the potential and kinetic energies. Since the mechanical energy is the total energy of the falling mass, the total mechanical energy should remain constant.

(f) The kinetic energy, the potential energy and the total energy (mechanical energy) will be plotted on a graph as a function of the distance fallen to show the variation of each during the free fall.

Procedure

1. The free fall apparatus should be properly leveled and wired prior to the start of the experiment. A power supply should be wired in series with a switch and the electromagnet of the Behr Free Fall Apparatus. The spark timer should not yet be plugged in and should be connected to the spark wires near the bottom of the apparatus. The spark timer switch should be open and the spark timer set at 60 Hz. The apparatus should have spark tape threaded through it.

2. Determine the mass of the free fall body in kilograms and record in Table 19 -1.

3. Energize the electromagnet at the top of the apparatus and suspend the mass from the electromagnet so that it hangs freely. With a pen or pencil mark the location of the metal ring that surrounds the free fall body on the spark timer tape. Label this mark point A.

CAUTION - HIGH VOLTAGE

THE SPARK TIMER IS TURNED ON IN STEP 4. A HIGH VOLTAGE IS PRESENT IN THE WIRES CONNECTED TO THE SPARK TIMER WHEN THE SPARK TIMER IS TURNED ON.

4. Plug in the spark timer. Push the "on" switch down for a second to be certain that there is a spark.

5. Having verified that the apparatus functions properly, hold the spark timer switch in the "on" position and open the switch to the electromagnet. This releases the suspended mass causing it to fall freely. Release the "on" switch of the spark timer when the body hits the bottom.

6. Unplug the spark timer.

7. Remove the strip of spark timer tape with the record of the free fall.

8. Examine the strip of paper and check that the spark timer made at least 21 successive spots after point A. If not, do the experiment over again up to this point.

9. Label each dot below point A with successive numbers ("1", "2", "3", etc.).

10. Measure the distances from dot A to each successive dot in meters (i.e., measure the distance s_1 from A to dot #1, the distance s_2 from A to dot #2, the distance s_3 from A to dot #3, etc.). See Figure 18-2 in the prior experiment.

11. Calculate and record in Table 19-1 the distances Δs between neighboring odd dots by subtracting the distances found in step 10 in the following way: $\Delta s_2 = s_3 - s_1$, $\Delta s_4 = s_5 - s_3$, $\Delta s_6 = s_7 - s_5$, etc. Each distance Δs is indicated in Table 19-1 by all even number because the Δs's will be used to determine the velocities of the freely falling body at the locations of the even numbered dots.

12. Calculate the average velocity between neighboring odd dots by dividing the Δs's by the time interval between the odd dots (1/30th of a second). Record these as the v's in Table 19-1. These average velocities are the actual velocities of the freely falling body at the even numbered dots.

13. Calculate the kinetic energies at the even numbered dots for your velocities from equation (19-2). Record your results in Table 19-1.

14. For each of your even numbered dots, calculate the decrease in potential energy relative to point A from equation (19-5). Record your values in Table 19-1.

15. Conservation of energy requires that the kinetic energy at each even numbered dot equal the decrease in the potential energy calculated in step 14. Using the decrease in the potential energy as your standard value, determine the percent error in your kinetic energy. Record your results in Table 19-1.

16. Conservation of energy also requires that the total energy of the system remains constant. In order to check this, do the following:

(a) Calculate and record in Table 19-1 the height above the 20th dot for each even numbered dot (i.e., $h_2 = s_{20} - s_2$, $h_4 = s_{20} - s_4$, $h_6 = s_{20} - s_6$, etc.)

(b) Calculate and record in Table 19-1 the potential energy relative to the 20th dot for each even numbered dot from equation (19-1).

(c) Calculate and record in Table 19-1 the total potential energy plus kinetic energy for each even numbered dot. This is the total energy of the freely fall mass at each even numbered point.

(d) The total energy from step 16(c) should be the same for all points. Does your data verify this? If so, all the values for the total energy should be the same (within experimental errors). If the total energy is constant, the kinetic energy should be increasing as the potential energy is decreasing. On a single sheet of graph paper draw a graph of each of the following (Be certain to label each line as to which type of energy it is):

(1) the kinetic energy as a function of the distance fallen, s,

(2) the potential energy as a function of the distance fallen, s, and

(3) the total energy as a function of the distance fallen, s.

Question

1. We used the distances between the odd numbered dots to calculate the Δs's from which the velocities were calculated. Why do we label the velocities with positive subscripts such as v_2 and v_4?

2. If energy is always conserved, what happens to the mechanical energy of the freely falling mass when it hits the ground?

EXPERIMENT 20

REACTION TIME (STATISTICS)

Objectives

1. To determine your average reaction time
2. To learn about statistics

Equipment

A meter stick

Discussion and Theory

The distance that an object will fall from rest is:

$$s = (1/2)\, g\, t^2 \qquad (20\text{-}1)$$

where t is the time in seconds and g, the acceleration due to gravity, is 9.8 m/s^2. If we solve for t we get:

$$t = (2s/g)^{1/2} \qquad (20\text{-}2)$$

In this experiment a meter stick will be dropped. The distance, s, it falls before it is caught will be used to determine your reaction time from equation (20-2). This procedure will be repeated many times and the results analyzed statistically.

In scientific work, when the same measurement is made repeatedly, instead of a single value, a range of values called a *distribution* is obtained. When dealing with distributions it is necessary to study them in an organized way. *Statistics* deals with the organization and analysis of numerical data.

Variations in data may be due to actual fluctuations in the quantity being measured and/or due to the limitations imposed by the measuring instrument. *Random variations* are due to actual variations of the measured quantity. They are beyond the control of the experimenter. *Systematic variations* are due to some irregularity that leads to a consistent error in all measurements of the quantity (e.g., measurements with a cloth tape measure that has stretched). This type of irregularity can be either corrected for or eliminated from the final results.

There are several ways of defining the "center" or *central tendency* of a distribution. These are:

(1) The *average value* or *mean value* is obtained by dividing the sum of all the values of the distribution by the total number of values.

(2) The *median* is the middle value of the distribution. There are just as many values above the median as there are below.

(3) The *mode* is the value that occurs most frequently.

The *standard deviation,* is used to measure the variability of the distribution. The standard deviation, σ, is defined as

$$\sigma = \sqrt{\frac{\Sigma (x_i - x_{ave})^2}{N - 1}} \qquad (20\text{-}3)$$

where Σ is the symbol for the sum

N = the total number of values in the distribution

x_i = the ith value of the distribution (where i goes from 1 to N)

x_{ave} = the average value of the distribution

A typical distribution curve for a large number of values with random errors is shown in Figure 20-1. Such a curve is called a *normal distribution curve.* In a normal distribution curve 68% of all values will fall between $x_{ave} + \sigma$ and $x_{ave} - \sigma$, and 95% of all values will fall between $x_{ave} + 2\sigma$ and $x_{ave} - 2\sigma$.

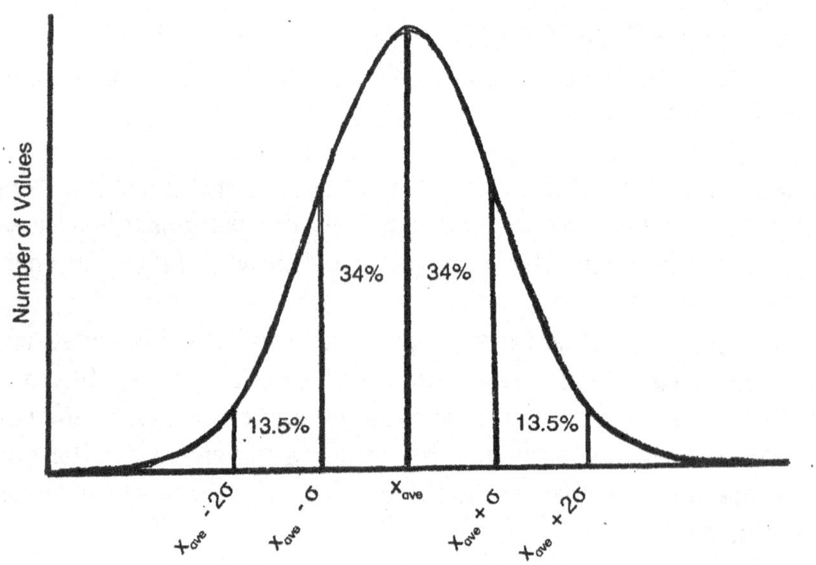

Figure 20-1 A Normal Distribution

Example

For the following data

Table 20-1

45	82	51	60	47	31	46	49	53	42	63	51
57	38	71	53	50	29	53	53	52	53	43	54

The average value is

$$x_{ave} = \frac{45+82+51+60+47+31+46+49+53+42+63+51+57+38+71+53+50+29+53+53+52+53+43+54}{24}$$

$$= 51.08 .$$

To find the median value organize the values in ascending order

Table 20-2

29	31	38	42	43	45	46	47	49	50	51	51
52	53	53	53	53	53	54	57	60	63	71	82

Since there are an even number of values, two values, 51 and 52, fall in the middle. The median is then taken to be half way between the two, i.e., 51.5.

To find the mode look for the number that appears most often in Table 20-2. In this case it is 53.

To find the standard deviation first calculate all the $(x_i - x_{ave})^2$ from Table 20-1:

$(45-51.08)^2=36.966$ $(82-51.08)^2=956.046$ $(51-51.08)^2=0.006$ $(60-51.08)^2=79.566$
$(47-51.08)^2=16.646$ $(31-51.08)^2=403.206$ $(46-51.08)^2=25.806$ $(49-51.08)^2=4.326$
$(53-51.08)^2=3.686$ $(42-51.08)^2=82.446$ $(63-51.08)^2=142.086$ $(51-51.08)^2=0.006$
$(57-51.08)^2=35.283$ $(38-51.08)^2=171.086$ $(71-51.08)^2=396.806$ $(53-51.08)^2=3.686$
$(50-51.08)^2=1.166$ $(29-51.08)^2=487.526$ $(53-51.08)^2=3.686$ $(53-51.08)^2=3.686$
$(52-51.08)^2=0.846$ $(53-51.08)^2=3.686$ $(43-51.08)^2=65.286$ $(54-51.08)^2=8.526$

Next sum all the $(x_i - x_{ave})^2$:

$$\Sigma (x_i - x_{ave})^2 = 36.966 + 956.046 + 0.006 + 79.566 + 16.646 + 403.206 + 25.806 + 4.326 + 3.686 + 82.446 + 142.086 + 0.006 + 35.283 + 171.086 + 396.806 + 3.686 + 1.166 + 487.526 + 3.686 + 3.686 + 0.846 + 3.686 + 65.286 + 8.526$$

$$= 2{,}932.061$$

Finally, Calculate the standard deviation from equation (20-3):

$$\sqrt{\frac{\Sigma (x_i - x_{ave})^2}{N - 1}} = \sqrt{\frac{2{,}932.061}{24-1}} = 11.3$$

If this is a normal distribution, 68% of all values should fall between (rounded to the nearest tenth)

$$x_{ave} - \sigma = 51.1 - 11.3 = 39.8 \text{ and } x_{ave} + \sigma = 51.1 + 11.3 = 62.4.$$

There are 18 values or 75% within this range.

For a normal distribution, 95% of all values should fall between

$$x_{ave} - 2\sigma = 51.1 - 2x11.3 = 28.5 \text{ and } x_{ave} + 2\sigma = 51.1 + 2x11.3 = 73.7.$$

There are 23 values or 96% within this range.

Figure 20-2 shows the distribution of these values in intervals of 5. Such a distribution, where the number of values that fall in an interval versus the interval, is called a *histogram*. In the first interval, 26 to 30, there is only one value. The same is true for the second and third intervals, 31 to 35 and 36 to 40. In the fourth interval, 41 to 45, there are three values, etc. The maximum number of values, 9, occurs in the sixth interval, 51 to 55.

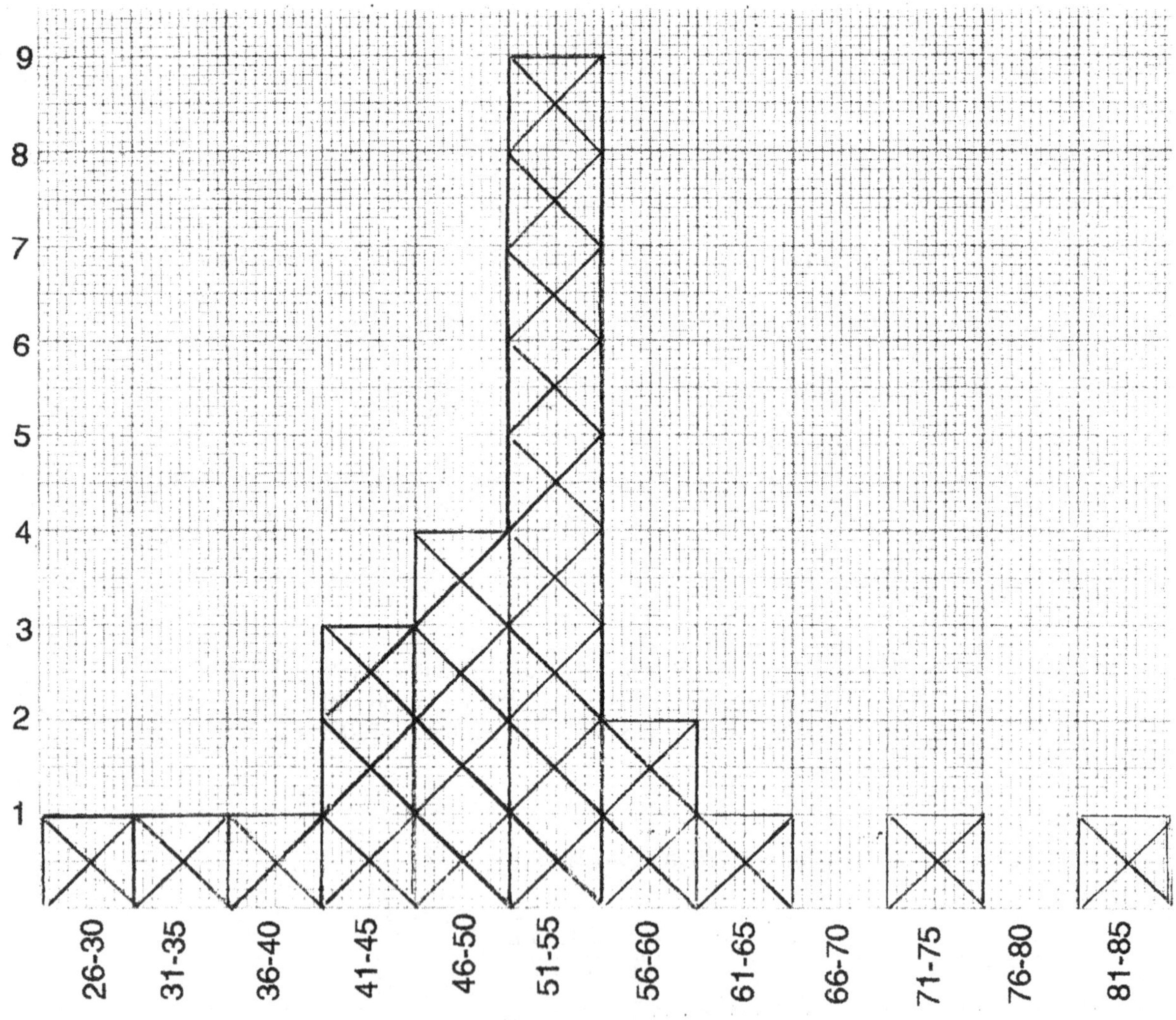

Figure 20-2 Histogram for Table 20-2

Procedure

1. Student A holds the meter stick so that the zero position is between the thumb and the first finger of student B. Student B keeps his fingers apart.

2. Student A drops the meter stick which student B catches with his two fingers at the instant he realizes the meter stick is falling.

3. Student B records in Table 20-4 the distance fallen by the stick.

4. The above procedure is repeated for as many times as directed by your instructor. Each time student B records the distance fallen in Table 20-4.

5. In Table 20-3 the time for an object to fall various distances is calculated from equation (20-2). Use Table 20-3 to determine the reaction time for each run. Record the results in Table 20-4.

Table 20-3
Time to Fall Various Distances

s (cm)	t (s)	s (cm)	t (s)	s (cm)	t (s)	s (cm)	t (s)	s (cm)	t (s)	s (cm)	t (s)
1	0.045	18	0.192	35	0.267	52	0.326	69	0.375	86	0.419
2	0.063	19	0.197	36	0.271	53	0.329	70	0.378	87	0.421
3	0.078	20	0.202	37	0.275	54	0.332	71	0.381	88	0.424
4	0.090	21	0.207	38	0.278	55	0.335	72	0.383	89	0.426
5	0.101	22	0.212	39	0.282	56	0.338	73	0.386	90	0.429
6	0.111	23	0.217	40	0.286	57	0.341	74	0.389	91	0.431
7	0.120	24	0.221	41	0.290	58	0.334	75	0.391	92	0.433
8	0.128	25	0.226	42	0.293	59	0.347	76	0.394	93	0.436
9	0.136	26	0.230	43	0.296	60	0.350	77	0.396	94	0.438
10	0.143	27	0.235	44	0.300	61	0.353	78	0.399	95	0.440
11	0.150	28	0.239	45	0.303	62	0.356	79	0.402	96	0.443
12	0.156	29	0.243	46	0.307	63	0.359	80	0.404	97	0.445
13	0.163	30	0.247	47	0.310	64	0.361	81	0.407	98	0.447
14	0.169	31	0.252	48	0.313	65	0.364	82	0.409	99	0.449
15	0.175	32	0.256	49	0.316	66	0.367	83	0.411	100	0.452
16	0.181	33	0.260	50	0.320	67	0.370	84	0.414	101	0.454
17	0.186	34	0.263	51	0.323	68	0.373	85	0.416	102	0.456

6. Calculate and record at the end of Table 20-4 the average, t_{ave}, of the reaction time;

$$t_{ave} = \frac{\Sigma t_i}{N}$$

where Σt_i = the sum of all values of the reaction time
N = the total number of values of the reaction time.

7. Calculate and record in Table 20-4 the deviation, D, from the average for each reaction time;

$$D = t_i - t_{ave}$$

where t_i = the reaction time of each individual run
t_{ave} = average reaction time calculated in step 6.

8. Calculate the square of each of the deviations, D^2. Record in Table 20-4

9. Find the sum of the squares of all the deviations, $\Sigma (D^2)$. Record your results at the end of Table 20-4.

10. Calculate and record at the end of Table 20-4 the standard deviation, σ, from

$$\sigma = \sqrt{\frac{\Sigma(D^2)}{N-1}}$$

where $\Sigma(D^2)$ = the sum of the squares of all the deviations calculated in step 9
N = the total number of values of the reaction time.

11. Determine $x_{ave} + \sigma$ and $x_{ave} - \sigma$, and $x_{ave} + 2\sigma$ and $x_{ave} - 2\sigma$. Record your values at the end of Table 20-4.

12. Determine how many values, N_σ, fall between $x_{ave} + \sigma$, and $x_{ave} - \sigma$ and how many values, $N_{2\sigma}$, between $x_{ave} + 2\sigma$ and $x_{ave} - 2\sigma$. Record your N_σ and $N_{2\sigma}$ at the end of Table 20-4.

13. Determine what percent of your values fall between $x_{ave} + \sigma$, and $x_{ave} - \sigma$ and between $x_{ave} + 2\sigma$ and $x_{ave} - 2\sigma$. Record your values at the end of Table 20-4.

Questions

1. Two researchers perform different experiments to determine the size of the same physical quantity. Both perform their experiment 100 times. Researcher A obtains an average value of 1.18 with a standard deviation of 0.13. Researcher B obtains an average value of 1.29 with a standard deviation of 0.04. Whose data is more variable?

2. What is the median value of your distribution?

3. Consider all values of the reaction time rounded to the nearest 0.1 seconds What is the mode of your distribution?

Name________________________________ Date Performed________________________________
Lab Partner________________________________ Date Submitted________________________________

Table 20-4
Statistical Analysis of Reaction Time

	d_i	t_i	$D = t_i - t_{ave}$	$D^2 = (t_i - t_{ave})^2$
1				
2				
3				
4				
5				
6				
7				
8				
9				
10				
11				
12				
13				
14				
15				
16				
17				
18				
19				
20				
21				
22				
23				
24				
25				
26				
27				
28				
29				
30				
31				
32				
33				
34				
35				
36				
37				
38				
39				
40				

Name________________________________

Table 20-4 (Continues)

	d_i	t_i	$D = t_i - t_{ave}$	$D^2 = (t_i - t_{ave})^2$
41				
42				
43				
44				
45				
46				
47				
48				
49				
50				
51				
52				
53				
54				
55				
56				
57				
58				
59				
60				
61				
62				
63				
64				
65				
66				
67				
68				
69				
70				
71				
72				
73				
74				
75				
76				
77				
78				
79				
80				

t_{ave} = ____________ $\Sigma(D^2) = \Sigma(t_i - t_{ave})^2$ = ____________ σ = ____________

$x_{ave} + \sigma$ = ____________ $x_{ave} - \sigma$ = ____________ N_σ = ____________ % within range = ____________

$x_{ave} + 2\sigma$ = ____________ $x_{ave} - 2\sigma$ = ____________ $N_{2\sigma}$ = ____________ % within range = ____________

EXPERIMENT 21

THE SPEED OF SOUND

Objective

To determine the speed of sound in air.

Equipment

Tuning forks, (with frequencies between 440 and 2000 Hz), resonance apparatus, striker, Celsius thermometer.

Discussion and Theory

Figure 21-1 shows a simple repetitive wave. The length, λ, at which the wave repeats itself is called the *wavelength*. If this wave is traveling with a constant velocity, v, the time it takes one wavelength to pass a fixed point is called the *period*. The number of waves per unit of time that will pass through a fixed point is called the *frequency*. The frequency, f, and the period, T, are reciprocals,

$$f = \frac{1}{T} \tag{21-1}$$

The speed, v, is defined as the distance, d, traveled per unit time, t,

$$v = \frac{d}{t} . \tag{21-2}$$

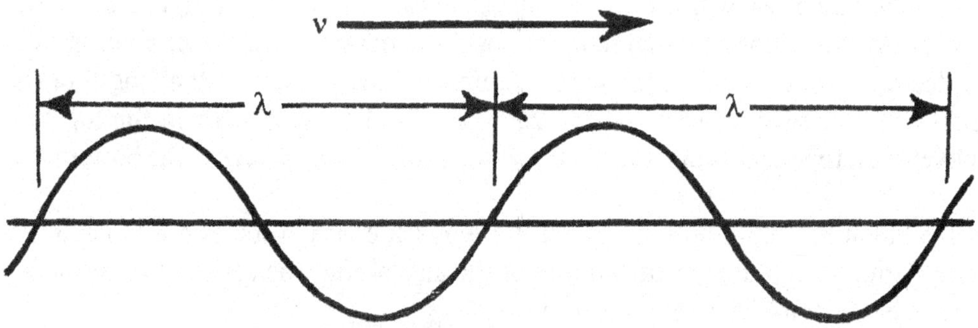

Figure 21-1 The Wavelength of a Repetitive Wave

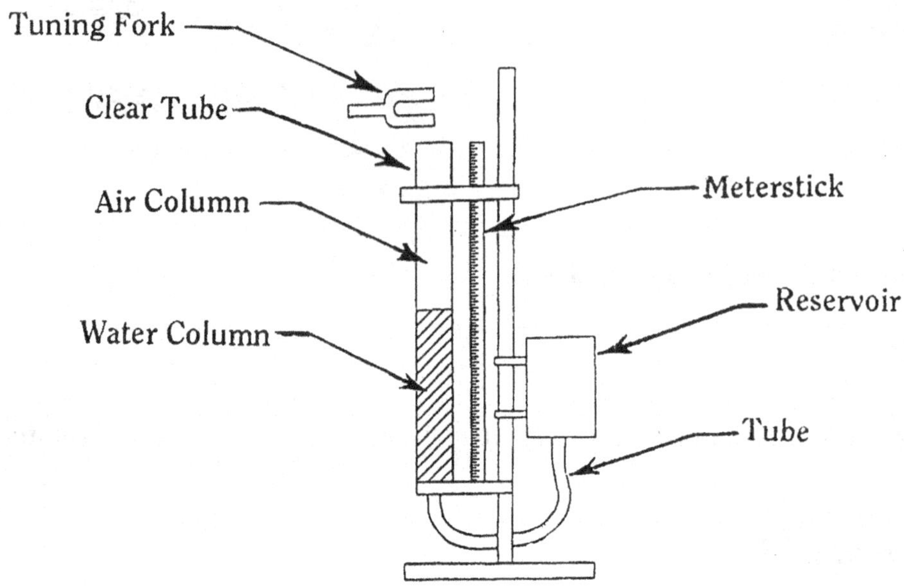

Figure 21-2 The Resonance Apparatus

For a wave to travel a distance of one wavelength, λ, the time is T, one period. Therefore,

$$v = \frac{\lambda}{T}. \tag{21-3}$$

Because 1/T = f, equation (21-3) becomes

$$v = \lambda f. \tag{21-4}$$

We will use equation (21-4) to obtain the speed of sound.

In this experiment we will use the resonance apparatus shown in Figure 21-2. The reservoir is filled with water. As the reservoir is moved up or down the height of the water column in the clear tube increases or decreases. The height of the water column is used to control the length of the air column between the top of the water column and the top of the clear tube. The air in the top part of the clear tube is effectively in a tube that is open at the top and closed (by the water) at the bottom.

When a vibrating tuning fork is held at the top of the clear tube a series of sound waves travels down the air column in the tube. At the bottom of the air column the series of sound waves is reflected back up the tube by the top of the water column.

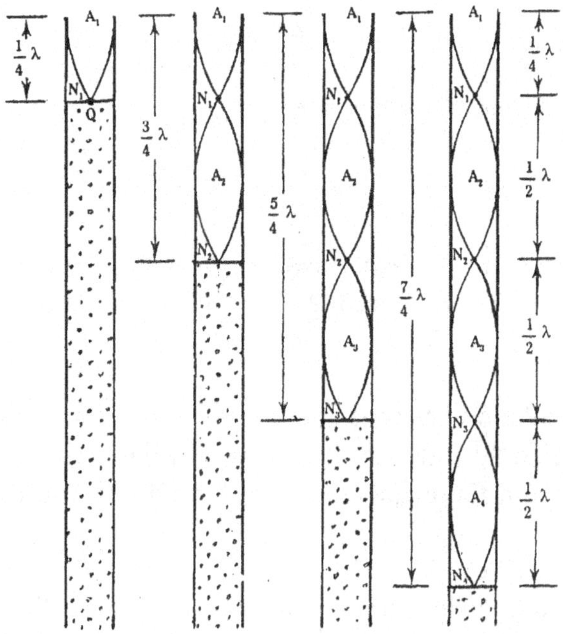

Figure 21-3 Standing Waves in an Air Column

If the water column starts at the top of the clear tube and is gradually lowered, the length of the air column increases. When the length of the air column equals about one-quarter of the wavelength of the sound coming from the tuning fork, a *standing wave* occurs in the air column. In a standing wave there are positions called *nodes* where the reflected wave cancels the wave moving down the tube. At the top of the water column, point Q in Figure 21-3(a), a node , N_1, occurs. At the top open end of the tube an *antinode* occurs where the reflected wave and the wave moving down the tube add together. Whenever standing waves occur the system is said to be in *resonance.* The antinode, A_1, at the top open end of the tube causes an increase in the intensity of the sound heard during resonance.

If the water column continues to be lowered, *standing waves will form each time the length of the air column is increased by one-half of a wavelength* (see Figures 21-3(a), (b), (c) and (d)). Each new standing wave is always one-half of a wavelength longer then the preceding standing wave and has an additional node and antinode. All nodes are separated by a distance of one-half of a wavelength. All antinodes are also one-half of a wavelength apart. During resonance, one antinode is always at the open end of the tube causing an increase in the sound intensity.

In this experiment a tuning fork with a known frequency, f, is struck and held at the top of the open tube of the resonance apparatus as shown in Figure 21-2. As the reservoir is slowly lowered, the length of the air column in the tube increases. The length L between resonances is determined and the wavelength is calculated from

$$\lambda = 2L. \qquad (21\text{-}5)$$

Finally the speed of sound is calculated from $v = \lambda f$.

Procedure

1. The resonance apparatus used for this experiment is shown in Figure 21-2. Lower the reservoir to its lowest position and fill it with water. Move the reservoir up so that the clear tube is filled to within a few centimeters of the top.

2. Strike a tuning fork and hold it over the top of the tube. Slowly move the reservoir down to lower the level of the water. A maximum in the intensity of the sound will be heard at several points, P_1, P_2, P_3, etc. along the length of tube. Record the frequency of the tuning fork and the positions P_1, P_2, P_3, etc. in Table 21-1.

3. Determine the distance between neighboring points, $P_2 - P_1$, $P_3 - P_2$, $P_4 - P_3$ etc. These differences are called L_1, L_2, L_3, etc. These lengths should be approximately the same. Record these values in Table 21-1

4. Take an average of the L values. Record this value as L_{ave} in Table 21-1.

5. The wavelength is twice this average length L_{ave}. Calculate and record the wavelength in Table 21-1. Don't forget units.

6. Use equation 21-4 to calculate the speed of sound.

7. Repeat steps 2 to 6 for as many tuning forks as requested by your instructor. Record the values in Table 21-1.

8. Determine room temperature with a Celsius thermometer and record. The standard value for the speed of sound is

$$V = (330 + .61\ T)\ \text{m/s} \qquad (21\text{-}6)$$

where T is in degrees Celsius. Calculate V from equation (21-6) and record at the end of Table 21-1.

9. Determine the percent error in each of your experimental values for the speed of sound and record in Table 21-1.

Questions

1. As sound passes from one material to another the speed of sound changes, but the frequency does not. What must happen to the wavelength?

2. If both ends of a tube are closed, standing waves are only possible with nodes at each end of the tube. If the wavelength is λ, then what is the shortest length for the tube that will result in a standing wave?

3. When is the speed of sound greater, during the summer or the winter?

4. Which sounds have longer wavelengths, high frequencies or low frequencies?

Name ______________________ Date Performed ______________________

Lab Partners ______________________ Date Submitted ______________________

____

Table 21-1
Determination of the Speed of Sound

Frequency of tuning fork (f) = ____________

P_1 = ______________

L_1 = ______________

P_2 = ______________

L_2 = ______________

P_3 = ______________

L_3 = ______________

P_4 = ______________

L_4 = ______________

P_5 = ______________

L_{ave} = ______________

λ = ______________

v = ______________

% error = ______________

Frequency of tuning fork (f) = ____________

P_1 = ______________

L_1 = ______________

P_2 = ______________

L_2 = ______________

P_3 = ______________

L_3 = ______________

P_4 = ______________

L_4 = ______________

P_5 = ______________

L_{ave} = ______________

λ = ______________

v = ______________

% error = ______________

Name________________________

Figure 1-1 (Continued)
Determination of the Speed of Sound

Frequency of tuning fork (f) = __________

P_1 = ____________

L_1 = ____________

P_2 = ____________

L_2 = ____________

P_3 = ____________

L_3 = ____________

P_4 = ____________

L_4 = ____________

P_5 = ____________

L_{ave} = ____________

λ = ____________

v = ____________

% error = ____________

Frequency of tuning fork (f) = __________

P_1 = ____________

L_1 = ____________

P_2 = ____________

L_2 = ____________

P_3 = ____________

L_3 = ____________

P_4 = ____________

L_4 = ____________

P_5 = ____________

L_{ave} = ____________

λ = ____________

v = ____________

% error = ____________

Temperature (T) = ____________

V = 330 + 0.61T = ____________

EXPERIMENT 22

ARCHIMEDES PRINCIPLE

Objective

To verify Archimedes Principle.

Equipment

Balance, masses, cord, an overflow cup, two beakers, a paper clip (large), a metal rod to support the scale.

Discussion and Theory

When an object is immersed in a liquid it experiences an upward force called the *buoyancy force*. *Archimedes Principle* tells us that, *the buoyancy force on an immersed object is equal to the weight of the liquid displaced.*

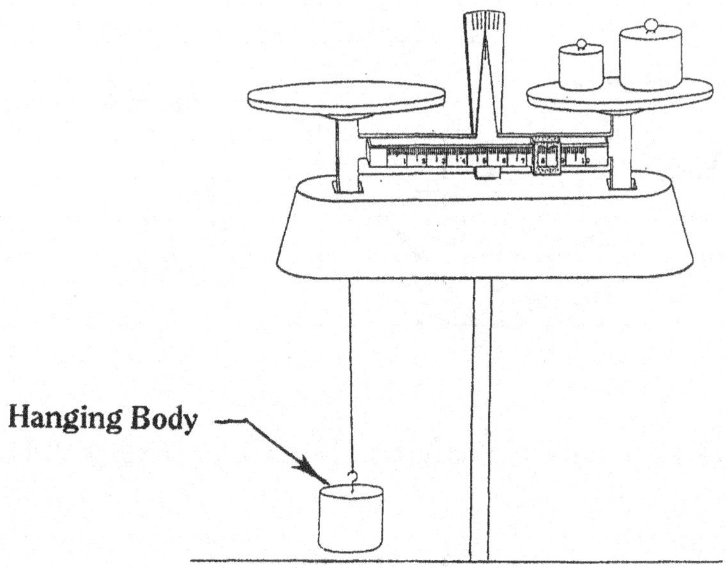

Figure 22-1 Determining the True Mass of the Hanging Body

In this experiment, we will verify Aechimedes Principle for objects that are fully immersed in water. We will weigh a body in air while it is hanging freely from the underside of a balance (see Figure 22-1). This weight is the true weight of the body. We will then lower the body into an overflow cup that is filled to the brim with water. The immersed body will displace some of the water causing the cup to overflow. We will capture the overflow and weigh it. We will also weigh the body as it hangs freely in the water from the underside of the balance (see Figure 22-2). This "weight" is the apparent weight of the body. The apparent decrease in the weight of the hanging body when it is immersed in water is due to the buoyancy force. Since the captured overflow water is equal to the amount of water displaced by the body, then the buoyancy force should equal the weight of this water.

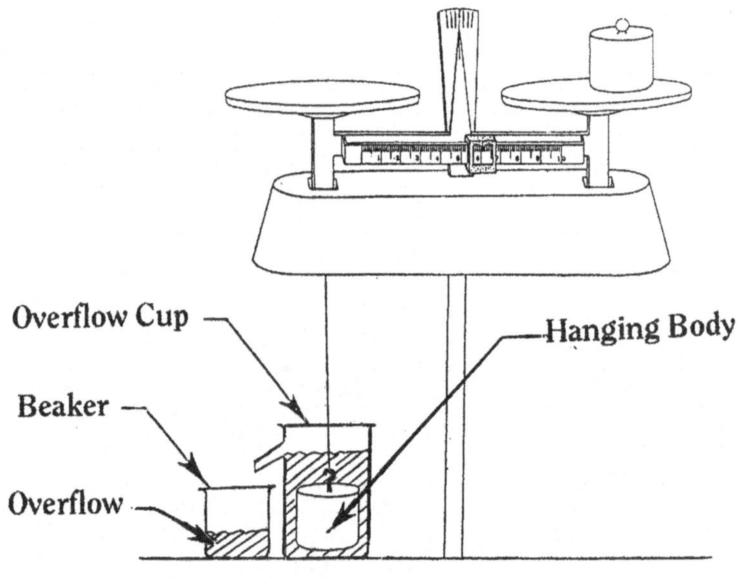

Figure 22-2 Determining the Apparent Mass of the Hanging Body

Procedure

1. Suspend a 0.200 kg (200 g) mass from the bottom of a balance as shown in Figure 22-1. Your instructor will show you how to attach it to the balance with some cord and a paper clip bent into an "S". This mass will to be used as the hanging body that will be immersed in water to verify Archimedes Principle.

2. Determine the total mass, M_T, of the hanging body in air (the mass itself plus the mass of the cord and paper clip) and record in Table 22-1.

3. Calculate the true weight, W_T, of the hanging body by multiplying M_T by 9.8 m/s^2.

4. Place the overflow cup under the balance as shown in Figure 22 -2. Place a beaker under the spout of the overflow cup. Fill the overflow cup with water until it overflows into the beaker. Let all the excess water overflow from the overflow cup into this beaker. Do not place the hanging body in the cup yet.

5. Determine the mass, M_B, of a dry beaker and record in Table 22-1.

6. Remove the beaker that captured the excess water. Place the dry beaker under the spout of the overflow cup. See Figure 22-2.

7. Slowly immerse the hanging body so that it hangs freely in the cup as shown in Figure 22-2. Be certain that the body does not touch the sides or rest on the bottom of the overflow cup.

8. Determine the apparent mass, M_A, of the hanging body immersed in water and record in Table 22-1.

9. Calculate the apparent weight, W_A, of the hanging body by multiplying the apparent mass, M_A, by 9.8 m/s^2. Record in Table 22-1.

10. Calculate the buoyancy force, F_B, by subtracting the apparent weight of the hanging body from the true weight, $W_T - W_A$. Record in Table 22-1.

11. Determine the mass, $M_{(B+W)}$, of the beaker with the overflow from the cup.

12. Determine the mass, M_W, of the water in the beaker by subtracting the mass of the dry beaker from the mass of the beaker with the overflow; $M_W = M_{(B+W)} - M_B$. Record in Table 22-1.

13. Determine the weight, W_W, of the overflow water by multiplying the mass M_W by 9.8 m/s^2 and record in Table 22-1.

14. The weight of the overflow water, W_W, should theoretically equal the buoyancy force, F_B. Calculate the percent error. Use the weight of the overflow water as your standard value.

12. Repeat steps 1 through 11 with a 0.500 kg (500g) mass used for your hanging body.

Questions

1. Why did we determine the masses of the 0.200 and 0.500 kg masses while suspended in air?

2. Cooking oil is "lighter then water" (i.e., it has a lower density). If the masses are lowered into cooking oil would the buoyancy force be more or less than when they are lowered into water?

3. Why must the hanging body be lowered slowly into the water?

Name________________________ Date Performed ________________________

Lab Partners ________________________ Date Submitted________________________

____

____

Table 22-1
Verification of Archimedes Principle

	0.200 kg Hanging Body	0.500 kg Hanging Body
True Mass of Hanging Body, M_T		
True Weight of Hanging Body, W_T		
Mass of Dry Beaker, M_B		
Apparent Mass of Hanging Body, M_A		
Apparent Weight of Hanging Body, W_A		
Buoyancy Force, F_B		
Mass of Beaker with Overflow, $M_{(B+W)}$		
Mass of Overflow, M_W		
Weight of Overflow, W_W		
Percent Error		

EXPERIMENT 23

COMPONENTS OF A VECTOR

Objective

To find the x and y components of a vector.

Equipment

Force table, weights, cord, mass hangers (4).

Discussion and Theory

A *vector* is a quantity that has magnitude (i.e. size) and direction. Any vector may be expressed as the sum of two perpendicular vectors called components. If a vector **F** is in a coordinate system as shown in Figure 23-1, then we will verify that the components of that vector may be expressed as:

$$F_x = F \cos \theta \quad (23\text{-}1)$$

$$F_y = F \sin \theta. \quad (23\text{-}2)$$

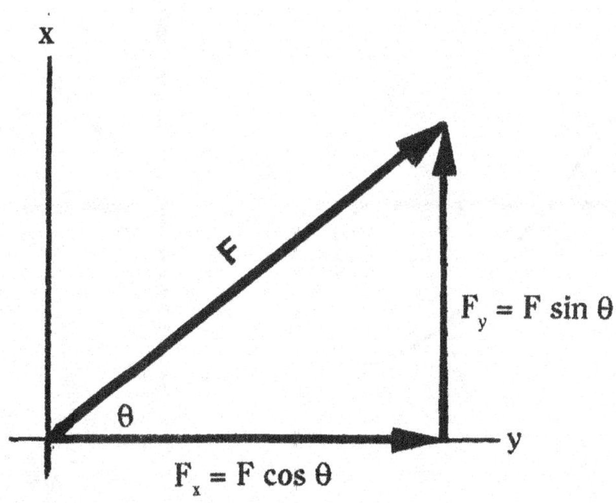

Figure 23-1 The Components of a Vector

Since F and its components form a right triangle, we should be able to show that the components satisfy the Pythagorean Theorem:

$$F^2 = F_x^2 + F_y^2 \tag{23-3}$$

and

$$F = \sqrt{F_x^2 + F_y^2} \tag{23-4}$$

In this experiment we will vary the direction of the vector F while keeping its magnitude constant. Because of equations (23-1) and (23-2) we should find that the graphs for the x and y components of **F** are as shown in Figure 23-2.

For the forces we will use the weights of three hanging masses. To obtain the weight W, multiply the mass m, by the acceleration due to gravity g

$$W = mg \tag{23-5}$$

If the mass is in kilograms then $g = 9.8\ m/s^2$ and the force is in Newtons.

(a) $F_x = F \cos \theta$ (b) $F_y = F \sin \theta$

Figure 23-2 Variation of the X and Y Components of a Vector

Procedure

1. Secure three movable pulleys on the force table; one at 0°, one at 180°, and one at 270°.

2. Secure the ring at the center of the force table with its retaining pin. Secure three cords to the retaining ring. Pass a cord from the retaining ring over each of the movable pulleys. Secure a mass hanger at the end of each cord.

3. Place 0.450 kg (450 grams) on a hanger at the 0° mark. Since the hanger is 0.050 kg (50 g), the total hanging mass M is now 0.500 kg. Record M in Table 23-1.

4. Calculate the weight of the mass M from equation (23-5) and record in Table 23-1. This weight is the applied force F on the ring. We are going to find the x and y components of the force needed to balance force F. Since these components are equal in size and opposite in direction to the components of force F, we will use them to determine the x and y components of F (F_x and F_y).

5. Add the proper mass at 180° to balnce the system. We will call this mass M_x since we will use its weight for F_x. You will find that the hanger at the 270° mark needs to be removed to obtain equilibrium. We will call the mass at 270°, M_y since we will use its weight for F_y. At this initial position, $M_y = 0$. Record M_x and M_y in Table 23-1. In doing this experiment you will be adding and removing masses to the hangers to obtain equilibrium. Before taking your readings, check the central ring onto which the cords are attached; the knots securing the cord should be positioned so that all the cords point towards the central hole as shown in Figure 23-3. When the system is in equilibrium the central hole of the ring should remain stationary over the retaining pin hole in the center of the force table when the retaining pin is removed.

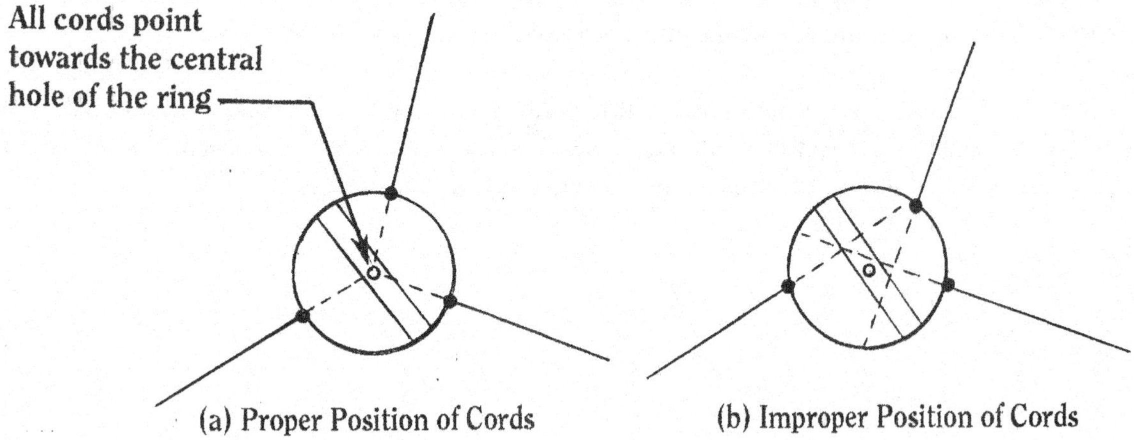

(a) Proper Position of Cords (b) Improper Position of Cords

Figure 23-3 Proper Position of Cords on Central Ring

6. Move the 0.500 kg body from the 0° to the 10° mark. Replace the hanger at the 270° mark. Add the proper mass at 180° and at 270° to balance the system. Record M_x and M_y in Table 23-1. (Be certain to include the 0.050 kg mass of the hangers in your values for M_x and M_y.

7. Move the 0.500 kg in 10° intervals up to 90°. At each 10° interval balance the system by adding or removing the appropriate masses at 180° and at 270°. Record M_x and M_y in Table 23-1 at each interval.

8. After 90°, move the 0.500 kg to 100°. To get the system to balance you will find that the pulley at 180° must be moved to the 0° mark. Moving the location of the hanger corresponds to a change in sign for the x-component of the force. Balance the system by adding and subtracting the appropriate masses at 0° and at 270°. Record M_x and M_y in Table 23-1.

9. Continue to move the 0.500 kg mass at 10° intervals until you reach 180°. Record M_x and M_y in Table 23-1 for each interval.

10. Calculate the weights of all the masses M_x and M_y. Record these weights as F_x and F_y in Table 23-1, since they are equal to the x and y components of the force F. These are your experimental values for the components of F.

11. Calculate and record the theoretical values of $F_x = F \cos \theta$ and $F_y = F \sin \theta$ for each of the indicated angles in Table 23-2. The force F is the weight of the 0.500 kg mass which was calculated in step 4.

12. Draw a smooth curve for $F \cos \theta$ (from Table 23-2) as a function of the angle. (Plot $F \cos \theta$ on the y-axis and the angles on the x-axis). Call this graph "F_x as a Function of Angle". On this graph, plot as x's your experimental values for F_x calculated in Table 23-1. Do not connect the x's. Do these points follow the curve? Discuss in your conclusion.

13. Draw a smooth curve for $F \sin \theta$ (from Table 23-2) as a function of the angle. Call this graph "F_y as a Function of Angle". On this graph, plot as x's your experimental values of F_y calculated in Table 23-1. Do not connect the x's. Do these points follow the curve? Discuss in your Conclusion.

14. The F_x's and F_y's in Table 23-1 are the x and y components of F. They should satisfy equation (23-4) at all intervals. Calculate $\sqrt{F_x^2 + F_y^2}$ at each interval. Calculate the percent error for each. The weight of the mass calculated in step 4 is your standard value. Discuss in your conclusion.

Questions

1. A force of 100 pounds makes an angle of 25° with the horizontal (the x-direction). Find the horizontal, F_x, and vertical, F_y, components of this force?

2. A force has an x-component of 50.0 lb and a y-component of 75.0 lb. What is the force?

3. From the results in Table 23-2, between what listed angles would you find $F_x = F_y$? What is the exact angle?

Name____________________ Date Performed____________________
Lab. Partners____________________ Date Submitted____________________
____
____

Table 23-1
The Experimental Components of a Vector

M = ____________________
F = Mg = ____________________

Location of the 0.500 kg	M_x (kg)	M_y (kg)	F_x (N)	F_y (N)	$\sqrt{F_x^2 + F_y^2}$ (N)	% Error
0	______	______	______	______	______	______
10	______	______	______	______	______	______
20	______	______	______	______	______	______
30	______	______	______	______	______	______
40	______	______	______	______	______	______
50	______	______	______	______	______	______
60	______	______	______	______	______	______
70	______	______	______	______	______	______
80	______	______	______	______	______	______
90	______	______	______	______	______	______
100	-______	-______	______	______	______	______
110	-______	-______	______	______	______	______
120	-______	-______	______	______	______	______
130	-______	-______	______	______	______	______
140	-______	-______	______	______	______	______
150	-______	-______	______	______	______	______
160	-______	-______	______	______	______	______
170	-______	-______	______	______	______	______
180	-______	-______	______	______	______	______

Name________________________________

Table 23-2
Theoretical Components of the Vector F

F = ________________

Angle (°) (°)	F cos θ (N)	F sin θ (N)
0		
10		
20		
30		
40		
50		
60		
70		
80		
90		
100		
110		
120		
130		
140		
150		
160		
170		
180		

EXPERIMENT 24

THE ADDITION OF VECTORS

Objective

To learn how to add forces.

Equipment

A force table, slotted masses, cord, mass hangers (4).

Discussion and Theory

In Experiment 23 *Resolution of Forces into Components,* we learned that a vector is a quantity that has both magnitude and direction. We also learned that a vector may be broken down into x and y components. In Figure 24-1 the vector **A** makes an angle θ with the x-axis. Notice that the vector **A** is written in bold lettering. Vectors will always be written in bold lettering to differentiate them from non-vectors (called scalars). The components of vector **A** are

$$A_x = A \cos \theta \qquad (24\text{-}1)$$

$$A_y = A \sin \theta. \qquad (24\text{-}2)$$

In this experiment we will learn how to add vectors together.

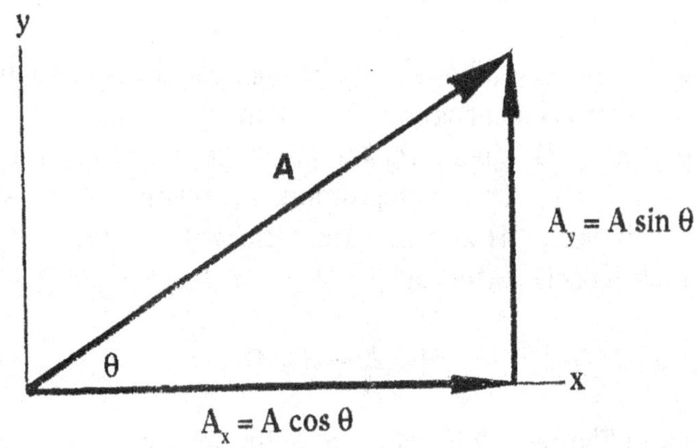

Figure 24-1 The x and y Components of a Vector

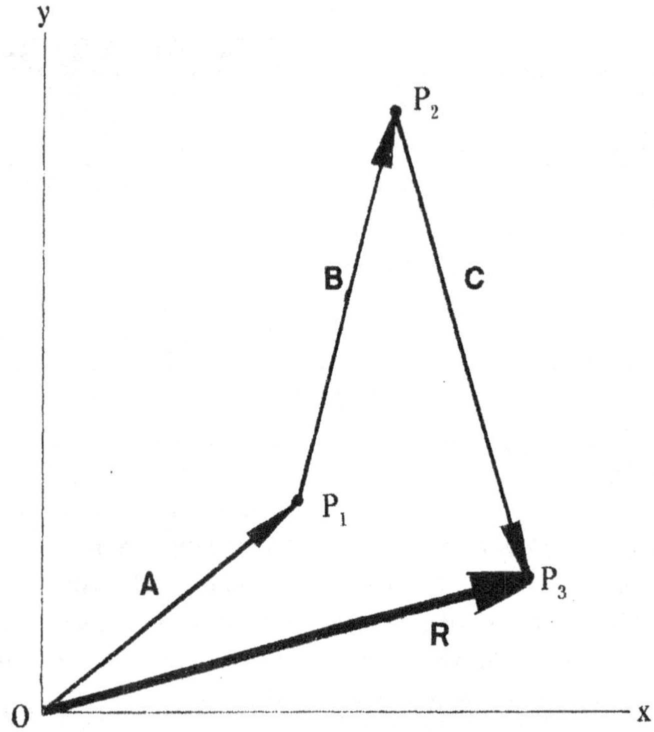

Figure 24-2 The Addition of Three Vectors

The *displacement* is defined as the change in position of an object. If a person traveled 25 kilometers from his home and then travelled an additional 10 kilometers, you don't know where he is unless you know the direction he travelled each time. In order *to add vectors, such as the displacement, it is always necessary to include the direction.*

In Figure 24-2 we see the path followed by a person. He started at point O and traveled along the straight path shown by the displacement vector **A** to point P_1. He then he travelled along the straight path depicted by the displacement **B** to point P_2, and finally he travelled along the straight path indicated by the displacement **C** to point P_3. The final position, P_3, relative to his starting position, O, is the displacement vector **R**. This vector **R** is called the *resultant*. It represents the effect of adding the vectors **A**, **B**, and **C** together. This is written as

$$\mathbf{R} = \mathbf{A} + \mathbf{B} + \mathbf{C}. \qquad (24\text{-}3)$$

Notice that the length of R in Figure 24-2 is not equal to the sum of the lengths A + B + C.

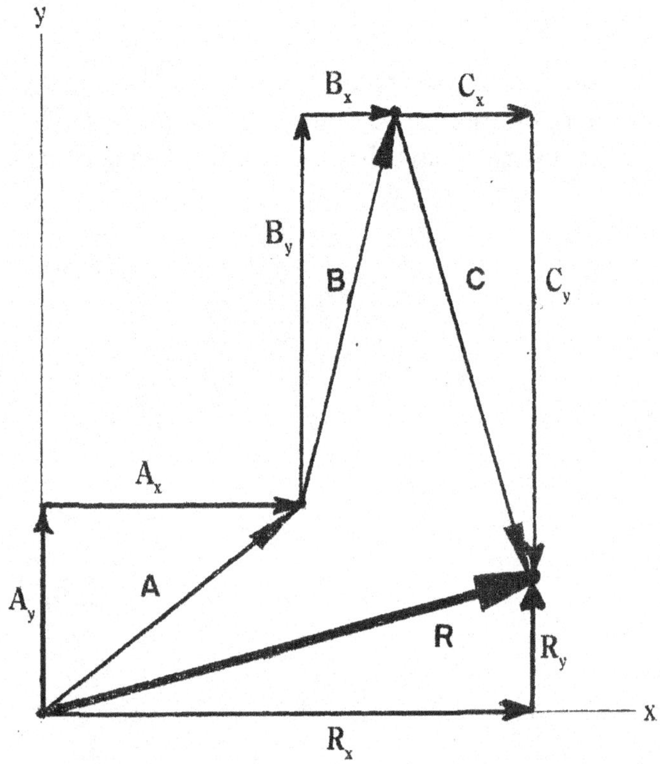

(a) The Components of the Vectors **A**, **B**, **C**, and **R**

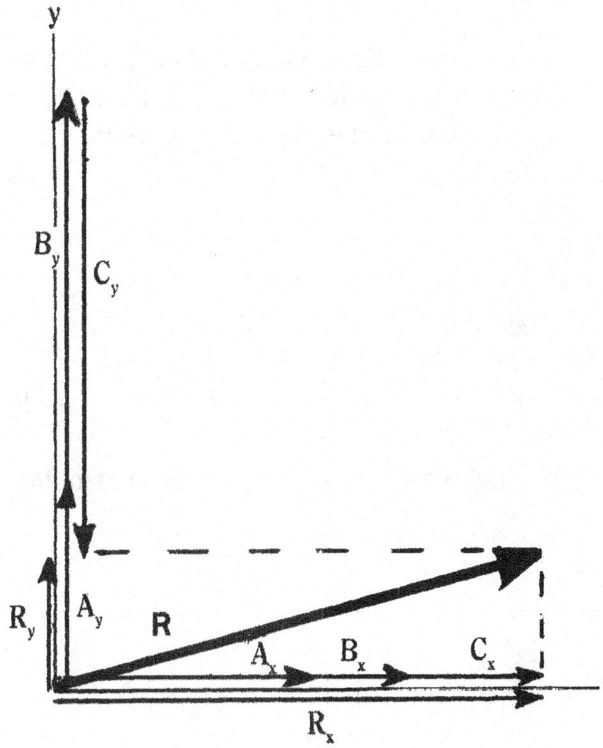

(b) Showing that R_x and R_y Equal the Sum of the Components

Figure 24-3 Finding R_x And R_y from the Components

Figure 24-3(a) shows the components of the vectors A, B, and C. In Figure 24-3(b) these same components are shown, but now with all the x-components on the x-axis and all the y-components on the y-axis. Notice that the length of R_x is the same as the sum of the components of A_x, B_x, and C_x;

$$R_x = A_x + B_x + C_x. \qquad (24\text{-}4)$$

Also notice, that the length of R_y is the same as the sum of the components of A_y, B_y, and C_y (since C_y points down it is treated as having a negative value that is added to the other components);

$$R_y = A_y + B_y + C_y \qquad (24\text{-}5)$$

Therefore, *the sum of the x-components of the vectors equals the x-component of the resultant, and the sum of the y-components of the vectors equals the y-component of the resultant*. In using equations (24-4) and (24-5) the student should be careful to include the proper signs when substituting for the components. In Figure 24-3(b) the y-component of vector C points down. To calculate R_y, a negative value must be substituted for C_y in equation (24-5).

Once the components of the resultant are known it is easy to find the resultant. The x and y components of the resultant and the resultant form a right triangle as shown in Figure 24-4. The magnitude of the resultant is found by applying the Pythagorean Theorem to the components. Therefore,

$$R^2 = R_x^2 + R_y^2 \qquad (24\text{-}6)$$

and

$$R = \sqrt{R_x^2 + R_y^2}. \qquad (24\text{-}7)$$

To find the direction we apply the tangent function

$$\text{Tan}\,\theta = \frac{R_y}{R_x} \qquad (24\text{-}8)$$

Therefore,

$$\theta = \text{Tan}^{-1}(R_y/R_x) \qquad (24\text{-}9)$$

In this experiment we will first determine a vector called the *equilibriant,* a vector equal in magnitude (size) to the resultant but in the opposite direction. The resultant will be determined from thr equilibriant.

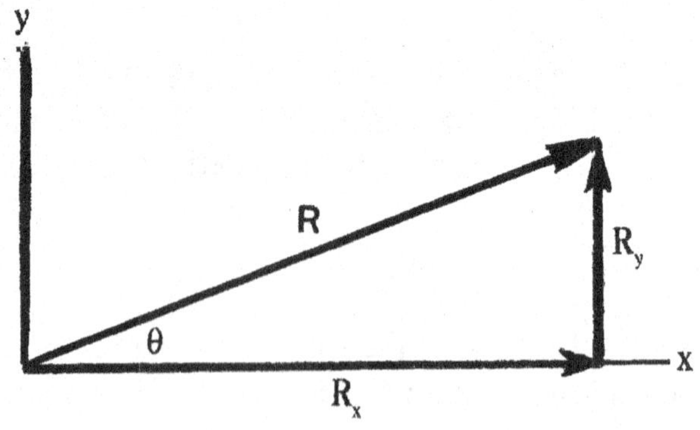

Figure 24-4 Finding the Resultant from R_x and R_y

Procedure

Add the following sets of vectors on the force table, unless different sets are given to you by your instructor:

Recall that the weight in Newtons is equal to mass in kilograms times 9.8 m/s^2 (i.e., W = mg) and that 1,000 g = 1.000 kg. All total hanging masses shown below include the mass of the hangers, 50 g.)

Set Number	Vector	Total Hanging Mass in grams	Total Hanging Mass in kilograms	Direction (°)	Force (N) (W = mg)
I	**A**	100	.100	0	0.980
	B	75	.075	90	0.735
	C	80	.080	135	0.784
II	**A**	100	.100	45	0.980
	B	150	.150	120	1.470
	C	80	.080	300	0.784
III	**A**	60	.060	90	0.588
	B	80	.080	115	0.784
	C	100	.100	180	0.980

For each set:

1. Secure the ring at the center of the force table with its retaining pin.

2. Position three of the movable pulleys at the indicated directions for the vectors to be added together. Position the fourth pulley at any location. Suspend four mass hangers at the ends of the cords and make certain that the cords pass over the pulleys. Add enough additional masses to the hangers to bring the total hanging mass up to the appropriate value for each.

3. Move the fourth pulley about the force table, and vary the total amount of hanging mass on it until equilibrium is attained. Check the central ring onto which the cords are attached; the knots securing the cord should be positioned so that all the cords point towards the central hole for the retaining pin as shown in Figure 23-3 in the prior experiment. When the system is in equilibrium the central hole of the ring should remain stationary over the retaining pin hole in the center of the force table. Record in Table 24-1 the direction of the fourth pulley and the total hanging mass for the equilibriant on your data sheet.

4. Convert the mass of the equilibriant to kilograms (1 kg = 1000 g). Multiply this value by 9.8 m/s^2 to obtain the equilibriant in Newtons. Record in Table 24-1

5. Determine and record in Table 24-1 your experimental resultant and its direction. The experimental resultant is equal in size to and opposite in direction to the equilibriant.

6. Calculate the x and y components of the given force vectors **A**, **B**, and **C** from equations (24-1) and (24-2). Record these values in Table 24-1. Be certain to include all signs

7. Calculate R_x and R_y from equations (24-4) and (24-5). Record in Table 24-1.

8. Determine the theoretical magnitude of the resultant from equation (24-7). Record in Table 24-1

9. Determine the angle, θ, the resultant makes with the positive x-axis from equation (24-9) and record in Table 24-1. Calculators cannot always be counted upon to give you the angle θ in the proper quadrant. If R_x and R_y are both positive the angle θ obtained from equation 24-9 is correct and in the first quadrant. If R_x and R_y are not both positive, first calculate the angle using equation 24-9 as if both R_x and R_y were positive. Next, let θ' represent the angle just calculated and find the correct angle, θ, as follows:

(a) If R_x is negative and R_y is positive the angle obtained is the number of degrees above the negative x-axis in the second quadrant. Subtract θ' from 180° to obtain θ.

(b) If both R_x and R_y are negative, the angle obtained is the number of degrees below the negative x-axis in the third quadrant. Add θ' to 180° to obtain θ.

(c) If R_x is positive and R_y is negative the angle obtained is the number of degrees below the positive x-axis in the fourth quadrant. Subtract θ' from 360° to obtain θ.

10. Use the value for the magnitude of the resultant obtained from step 8 as your standard. Determine and record the percent error in the experimental magnitude of the resultant.

11. Determine and record the difference between your experimental direction for the resultant and the theoretical value (from step 9)

Name________________________________ Date Performed________________________

Lab Partners ________________________ Date Submitted________________________

____

____

Table 24-1
The Addition of Vectors

Set Number I

Mass of equilibriant = ____________________ Direction of equilibriant = ______________

Magnitude of equilibriant = ________________

Magnitude of experimental resultant = __________ Direction of experimental resultant = ______

A_x = ______________________________ A_y = ______________________________

B_x = ______________________________ B_y = ______________________________

C_x = ______________________________ C_y = ______________________________

R_x = ______________________________ R_y = ______________________________

Theoretical magnitude of resultant = ___________ % error (experimental resultant) = ________

Theoretical direction of resultant = ____________ Difference between the experimental and theoretical directions = ____________

Set Number II

Mass of equilibriant = ____________________ Direction of equilibriant = ______________

Magnitude of equilibriant = ________________

Magnitude of experimental resultant = __________ Direction of experimental resultant = ______

A_x = ______________________________ A_y = ______________________________

B_x = ______________________________ B_y = ______________________________

C_x = ______________________________ C_y = ______________________________

R_x = ______________________________ R_y = ______________________________

Theoretical magnitude of resultant = ___________ % error (experimental resultant) = ________

Theoretical direction of resultant = ____________ Difference between the experimental and theoretical directions = ____________

Name________________________________

Table 24-1 (Continued)
The Addition of Vectors

Set Number III

Mass of equilibriant = ________________________

Direction of equilibriant = ________________

Magnitude of equilibriant = __________________

Magnitude of experimental resultant = __________

Direction of experimental resultant = ______

A_x = ________________________________

A_y = ________________________________

B_x = ________________________________

B_y = ________________________________

C_x = ________________________________

C_y = ________________________________

R_x = ________________________________

R_y = ________________________________

Theoretical magnitude of resultant = ____________

% error (experimental resultant) = ________

Theoretical direction of resultant = ____________

Difference between the experimental and theoretical directions = ____________

EXPERIMENT 25

LINEAR THERMAL EXPANSION

Objective

To study the linear thermal expansion of metals.

Equipment

Linear expansion apparatus, metal rods, thermometer, steam generator, Bunsen burner, meter stick, rubber tube, power supply.

Discussion and Theory

When a metal rod is heated, its length increases. This is called *linear thermal expansion*. The increase in the length of the rod is given by the relationship:

$$\Delta L = \alpha\ L_0 \Delta T \tag{25-1}$$

where

ΔL = the change in length of the rod
L_0 = the initial length of the rod
ΔT = the temperature change
α = the coefficient of linear thermal expansion.

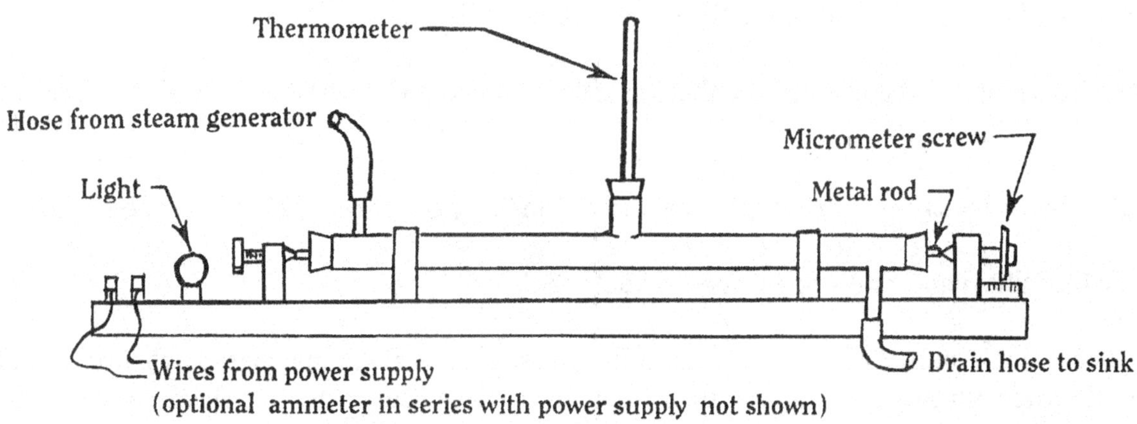

Figure 25-1 The Linear Expansion Apparatus

In this experiment, the coefficient of linear thermal expansion for one or more metals will be determined with the Linear Thermal Expansion Apparatus shown in Figure 25-1. A metal rod is held horizontally through the center of the apparatus by two stoppers at each end of the rod. A thermometer is held in position in the top of the apparatus by a stopper. Power from a power supply goes through a series circuit that passes through a light bulb on the apparatus and through the rod by means of two thumb screws when they make contact with the ends of the rod. One of the thumb screws is a micrometer screw which is used to measure the increase in length of the rod when it is heated. The light is used to detect the contact made by the micrometer with the rod. To heat the rod, steam from a steam generator is supplied to the apparatus by means of a hose. Condensed water and excess steam leave the apparatus by means of a drain hose. Often an ammeter is installed in series with the apparatus in the circuit from the power supply. It is found that the ammeter responds more quickly than the light when electrical contact is made.

Table 25-1
Coefficients of Linear Expansion

Metal	α (1 / °C)
Aluminum	2.4×10^{-5}
Brass	1.95×10^{-5}
Copper	1.7×10^{-5}
Steel	1.15×10^{-5}

Procedure

1. Measure the length, L_o, of the metal rod. Record this length and the metal from which the rod is made in Table 25 -2.

2. Insert the metal rod into the long tube of the linear thermal expansion apparatus. Be certain that the rod has passed through the hole in the stopper at each end of the tube.

3. Gently position the rod between the thumb screw and micrometer screw so that the rod just touches both.

4. Pass a thermometer through a stopper and insert it in the space provided at the top of the apparatus.

5. Read the initial temperature and record in Table 25-2.

6. Connect the terminals of the apparatus to the power supply. If an ammeter is supplied, install the meter in series with the apparatus and the power supply. Have your instructor check the circuit before turning on the power.

7. Your instructor will explain the use of the micrometer screw at one end of the apparatus. When the power is turned on, back off the micrometer screw so that there is no electrical contact. Slowly turn the micrometer screw until the light or ammeter indicates electrical contact is made. Immediately stop turning the micrometer. Record the micrometer screw reading. Repeat this procedure four more times until five micrometer readings are obtained. Calculate the average of these values and record in Table 25-2.

8. Back off the micrometer two full turns.

9. Fill the steam generator half to two-thirds full with water. Connect the generator to the expansion apparatus by attaching a rubber hose from the apparatus to the steam generator. Connect the drain hose to the apparatus and let it drain into a sink.

10. Light a Bunsen burner below the steam generator. Wait until steam passes through the entire expansion apparatus and comes out of the drain hose. The thermometer should go up to nearly 100 °C. This temperature should be maintained for about five minutes . Read the temperature and record in Table 25-2.

11. Take five final micrometer screw readings by slowly tightening the micrometer screw until contact is indicated by the light or ammeter. Record these values and their average in Table 25-2.

12. Determine the change in length, ΔL, of the rod from the difference between the average final and average initial micrometer screw readings. If the micrometer goes through zero the final average micrometer reading will appear to be smaller than the initial micrometer reading. To obtain the correct change in length, add 1.00 mm to the final average micrometer reading before subtracting the initial average reading. Record the change in length in Table 25-2.

13. Calculate the change in temperature, ΔT, by subtracting the initial temperature from the final temperature and record in Table 25-2.

14. Determine the coefficient of linear expansion from

$$\alpha = \frac{\Delta L}{L_0 \, \Delta T} \qquad (25\text{-}2)$$

Record your result in Table 25-2.

15. Compare this value with the standard value of the coefficient of linear expansion and find the percent error.

16. Repeat steps 1 through 15 for all the rods supplied to you in the laboratory.

Questions

1. Of the metals listed in Table 25-1, (a) which expands the most when heated? (b) which expands the least?

2. How would you expect the value for the coefficient of linear expansion to be affected if (a) you used a rod that was twice as long, (b) the change in temperature doubled?

3. By how much would the length of a 1.00 meter long rod change if the temperature increases by 100 °C and the rod is made of (a) steel, (b) brass, (c) aluminium, (d) copper.

Name ______________________ Date Performed ______________________
Lab Partners ______________________ Date Submitted ______________________
____
____

Table 25-2
Determination of the Coefficient of Linear Expansion

Initial Micrometer Readings	Final Micrometer Readings

Average Initial
Micrometer Reading: ______________________

Average Final
Micrometer Reading: ______________________

Metal of Rod ______________________
Initial Length of Rod (L_0) ______________________
Initial Temperature ______________________
Final Temperature ______________________
Change in Temperature (ΔT) ______________________
Change in Length (ΔL) ______________________
Coefficient of Linear Expansion (experimental): ______________________
Coefficient of Linear Expansion (standard): ______________________
Percent Error: ______________________

Name____________________

Table 25-2 (Continued)
Determination of the Coefficient of Linear Expansion

Initial Micrometer Readings	Final Micrometer Readings

Average Initial
Micrometer Reading: ____________________

Average Final
Micrometer Reading: ____________________

Metal of Rod ____________________

Initial Length of Rod (L_0) ____________________

Initial Temperature ____________________

Final Temperature ____________________

Change in Temperature (ΔT) ____________________

Change in Length (ΔL) ____________________

Coefficient of Linear Expansion (experimental): ____________________

Coefficient of Linear Expansion (standard): ____________________

Percent Error: ____________________

EXPERIMENT 26

TEMPERATURE CONVERSION

Objectives

1. To determine the relationships for converting between the Fahrenheit and Celsius temperature scales.

2. To analyze the graph of a straight line with a non-zero y-intercept and to obtain the equation for such a line.

Equipment

A Fahrenheit thermometer with a range of 32 °F to 212 °F, a Celsius thermometer with a range of 0 °C to 100 °C, a Bunsen burner, a beaker, a beaker stand, tap water, ice.

Discussion and Theory

The student should review the discussion in Experiment 4, *Linear Graphs* prior to doing this experiment.

Two commonly used temperature scales are the Fahrenheit and Celsius scales. The standard equation for converting the temperature in degrees Celsius to the temperature in degrees Fahrenheit is

$$F = (9/5)\,C + 32 \qquad (26\text{-}1)$$

or

$$F = 1.80\,C + 32 \qquad (26\text{-}2)$$

where

F = the temperature in degrees Fahrenheit

C = the temperature in degrees Celsius

Equation (26-2) is of the form $y = mx + b$, which is the equation for a straight line. The slope, m (with units), of this line is 1.80 °F/°C and the y-intercept, b (with units), is 32 °F.

In this experiment you will make temperature measurements in degrees Celsius and degrees Fahrenheit and plot the temperature in degrees Fahrenheit as a function of the temperature in degrees Celsius. From the slope and y-intercept of your best straight line you will determine the equation of the line, your conversion equation, by substituting into

$$F = (\text{ the slope of your line}) C + (\text{ the y-intercept of your line}). \qquad (26\text{-}3)$$

Example

John determines the slope of his line to be 1.87 °F/°C and the y-intercept to be 34 °F, then his conversion equation from equation (26-3) is $F = 1.87C + 34$.

Solving equation (26-1) for the temperature in degrees Celsius yields

$$C = (5/9)\,(F - 32) \qquad (26\text{-}4)$$

or

$$C = 0.556\,(F - 32) \qquad (26\text{-}5)$$

Equation (26-5) converts the temperature in degrees Fahrenheit to the temperature in degrees Celsius.

To obtain the conversion from degrees Fahrenheit to degrees Celsius from your results substitute into

$$C = (\text{ value of the reciprocal of the slope})(F - [\text{ the y-intercept of your line }]) \qquad (26\text{-}6)$$

Example

John determines the reciprocal of his slope to be (1/1.87 °F/°C) = 0.535 °C/°F. Therefore, his conversion equation from equation (26-6) is $C = 0.537(F - 34)$.

Procedure

1. Measure the temperature of each of the following simultaneously in degrees Fahrenheit and degrees Celsius and record these values in Table 26-1.
   a) Room temperature
   b) Ice water
   c) Tap water

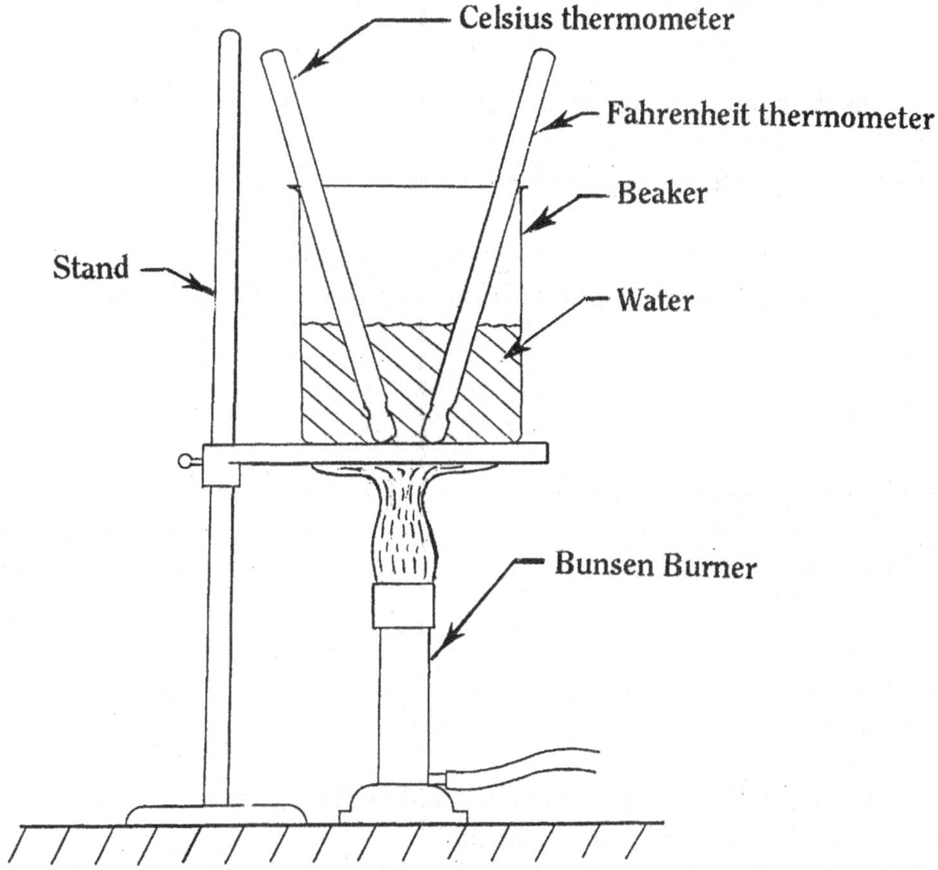

Figure 26-1 Heating Water while Measuring the Temperature with Celsius and Fahrenheit Thermometers

2. Set up a beaker stand, a beaker about half filled with water and a Bunsen burner as shown in Figure 26-1. Heat the water and record the temperature in degrees Fahrenheit in Table 26-1 as the water reaches each of the following temperatures in degrees Celsius: 30 °C, 40 °C, 50 °C, 60 °C, 70 °C, 80 °C, 90 °C and 100 °C.

3. Construct a graph of your data. Plot the temperature in degrees Celsius on the x-axis and the temperature in degrees Fahrenheit on the y-axis. Review *Setting up a Graph* and *Plotting the Data and Drawing the Straight Line* in the discussion of Experiment 4 if necessary.

4. Draw the best straight line for your data. Review *Plotting the Data and Drawing the Straight Line* in the discussion of Experiment 4 if necessary.

5. Determine the slope of your line. Review *The Analysis of a Straight Line* in the discussion of Experiment 4 if necessary.

6. Find the percent error for your slope and record in Table 26-1.

7. Determine the y-intercept of your line. Record in Table 26-1.

8. Find the percent error for the y-intercept and record in Table 26-1.

9. Write your experimental conversion equations from degrees Fahrenheit to degrees Celsius and from degrees Celsius to degrees Fahrenheit in the spaces provided on the data sheet

Questions

1. Which of the following are equations of a straight line:

(a) $y = 3x^2 + 5$
(b) $y = 0.12x + 0.89$
(c) $N = -5M + 168$
(d) $y = 8 + 2x$
(e) $y^2 = 2x + 3$

2. Can equation (26-5) be written in the standard form for a straight line (i.e. can it be written in the form $C = mF + b$)? If so, what is the equation?

3. Convert to degrees Fahrenheit:

(a) 85 °C
(b) 32 °C
(c) 165 °C
(d) -273 °C
(e) -40 °C

4. Convert to degrees Celsius:

(a) 98.6 °F
(b) 543 °F
(c) -32 °F
(d) 43 °F
(e) -100 °F

EXPERIMENT 27

RATES OF COOLING (DECAY PROCESSES)

Objective

1. To study the relative cooling rates of objects:
   (a) When they are in direct contact with the surrounding air
   (b) When there is an air gap between the object and the surrounding air
   (c) When styrofoam insulation surrounds the object
2. To determine the decay constant for each of the above cases.
3. To better understand decay processes.

Equipment

Two aluminum calorimeters (complete assemblies less stirrer), 1 additional inner aluminum calorimeter cup, 1 additional calorimeter cover, 1 styrofoam calorimeter cup with top plastic ring (Pasco Scientific part # 740-02975), 3 thermometers, 3 single hole stoppers for the calorimeter covers, 1 paper or styrofoam coffee cup (marked so that the inner calorimeter cups will fill to about 3/4 capacity), hot water.

Discussion and Theory

The rate of cooling of an object is affected by how well it is insulated from the surrounding air. In this experiment, we will compare the rates of cooling for the following three cases:

1. A totally uninsulated covered cup filled with hot water (Figure 27-1).

2. A cup filled with hot water in a calorimeter so that there is an air gap between the inner and outer cups of the calorimeter (Figure 27-2).

3. A cup filled with hot water in a calorimeter with styrofoam insulation in the gap between the inner and outer cups of the calorimeter (Figure 27-3).

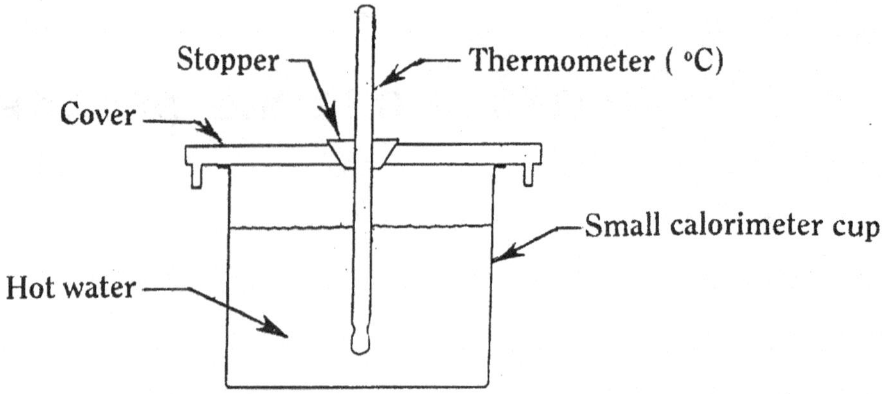

Figure 27-1 Totally Uninsulated Cup

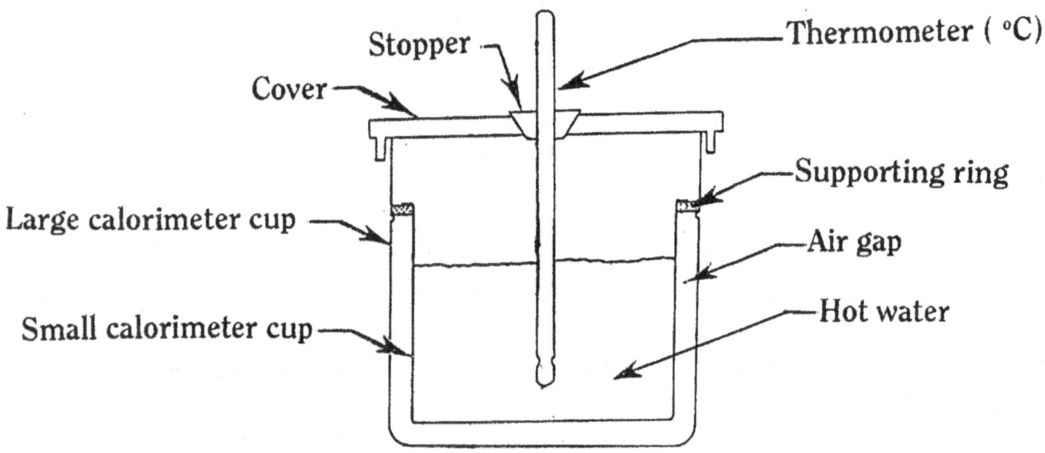

Figure 27-2 Cup with an Air Gap

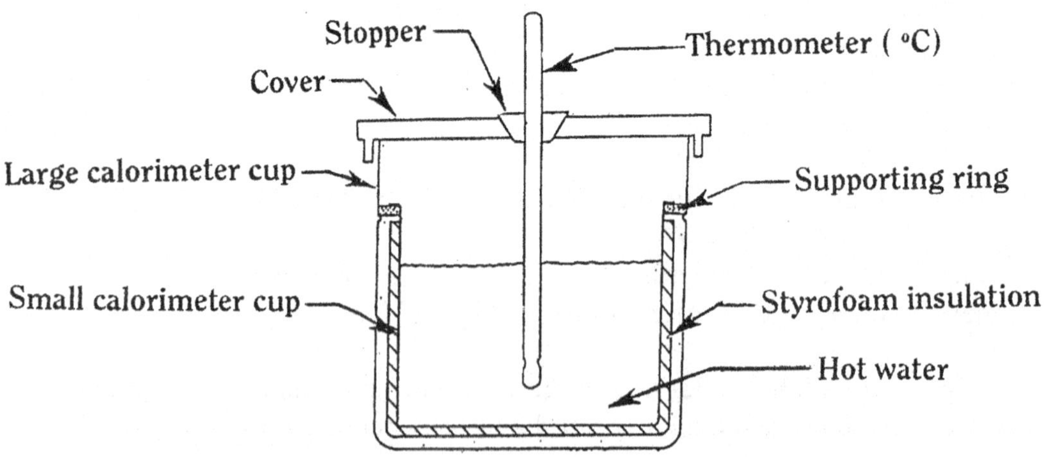

Figure 27-3 Styroform Insulated Cup

In any of the above cases, if the hot water is initially at a temperature T_O, and the temperature of the surrounding air is T_R, then at any time, t, the temperature of the hot water exceeds the temperature of the surrounding air by an amount, ΔT, given by the following:

$$\Delta T = (T_O - T_R)e^{-\lambda t} \tag{27-1}$$

where

λ = the decay constant
e = 2.718 (rounded to 4 significant figures).

The actual temperature of the water, T, at any time is

$$T = T_R + \Delta T$$

$$T = T_R + (T_O - T_R)\, e^{-\lambda t} \tag{27-2}$$

All naturally occurring decay processes are observed to involve the quantity "e". This quantity is called the *base of the natural* logarithm. In natural decay processes it always occurs raised to a negative power of some constant multiplied by time as it does in equation (27-1). Therefore, naturally occurring decay processes, such as radioactive decay, behave in a manner similar to the cooling process.

Procedure

1. Use one of the supplied thermometers to measure room temperature T_R. This is the temperature of the air surrounding each of the three objects we will consider (see below). Record the room temperature in Table 27-1.

2. Prepare three small cups as shown in Figures 27-1, 27-2 and 27-3 so that they are ready to be filled with hot water. Note that:

(a) In Figure 27-1 (uninsulated case), there is a small cup with a cover assembly which consists of a cover, a stopper in the hole on the cover and a thermometer through the hole in the stopper.

(b) In Figure 27-2 (air gap case), the small cup is placed in the larger cup of the calorimeter and is supported by a plastic supporting ring. The cover assembly is similar to that in (a) above.

(c) In Figure 27-3 (styrofoam insulated case), the aluminum calorimeter and cover are as in (b) except that the small aluminum calorimeter cup is placed within a styrofoam cup with a plastic ring to hold it in place before being placed within the large aluminum calorimeter cup.

3. A coffee cup has been marked on the inner surface so that equal amounts of hot water can be measured. Use this cup to fill each of three small calorimeter cups with equal amounts of hot water. Work quickly so that all three calorimeter cups will have the same starting temperature Be certain that the thermometers are pushed all the way down into the water.

4. As quickly as possible, read and record in Table 27-1 the temperature, T, of the water at five-minute intervals for 90 minutes for each case.

5. On a single sheet of linear graph paper, plot the temperature, T, as a function of time for each case. Draw the curve for each case and label each of the three curves. Call this Graph # 1.

6. For all the temperatures obtained in each case, calculate the natural logarithm and record it in the column labeled "Ln T" in Table 27-1. To find the natural logarithm on most simple scientific calculators, first punch-in the number and then push the button labeled "Ln x" or "Ln". For programmable calculators punch "Ln", then the number followed by "ENTER"

Example

To find the natural logarithm of 94
(a) on a simple calculator, punch in "94", then "Ln" to obtain 4.543....
(b) on a programmable calculator, push "Ln" then punch in "94" followed by "ENTER" to obtain 4.543....

7. On a separate sheet of graph paper for each of the three cases, plot the elapsed time in seconds on the x-axis and the values you calculated for LnT on the y-axis. Call these graphs "Graph 2, Uninsulated Case", "Graph 2, Air gap Case", and "Graph 2, Insulated Case". Analyze each graph as follows:

a. Draw the best straight line for each graph. Your **line should tilt downward.**

b. Calculate the slope of the line. Show the calculation on the graph. You should obtain a negative number. The slope is the negative of the decay constant (i.e., it is $-\lambda$). Record the decay constant in Table 27-1 for each case.

8. On the reverse side of your data sheet (page 188) you are asked to write the equation for the temperature for each case. To do this, do the following:

(a) Calculate $(T_O - T_R)$ I where To is the temperature at t = 0 seconds.

(b) Substitute the values for T_O, $(T_O - T_R)$ and $-\lambda$for each case into equation (27-2) and write the equation in the space provided.

Example

For one of the cases, a student obtained the results $T_O = 90\ °C$, $T_R = 21°C$ and $-\lambda = -1.25 \times 10^{-4}/s$. Therefore, $T_O - T_R = 90 - 21 = 69\ °C$. The correct equation for the temperature is found by substituting into

$$T = T_R + (T_O - T_R)\, e^{-\lambda t}$$

to obtain

$$T = 21 + 69e^{-1.25\times10^{-4}t}.$$

Questions

1. Examine Graph 1. In which case is cooling the fastest? For which case is it the slowest? Is this what you expected?

2. Examine the rate at which the temperature drops during the full 90 minutes for each case in Graph 1. Does the rate of temperature drop seem to increase or decrease during cooling?

3. Thermal windows have an air gap between two pieces of glass to reduce heat loss. On the basis of the results of this experiment, does an air gap appear to reduce heat loss effectively?

4. The half-life is the time needed for a radioactive substance to undergo radioactive decay so that only half of the initial amount remains. The cooling of hot water to room temperature is analogous to radioactive decay with the "half-life" equal to the time it takes the water to cool half-way to room temperature. For each case in this experiment (a) what is $(T_O - T_R)/2$ and (b) what is the "half-life", as read from Graph 1 for the initial temperature of the water to drop by $(T_O - T_R)/2$.

5. If the half-life (see question 4) of a radioactive substance is 10.0 minutes and there were initially 1,000,000 atoms of that substance, (a) how many would remain after 10.0 minutes? (b) after 20.0 minutes? (c) after 40 minutes?

Name ______________________ Date Performed ______________________

Lab Partners ______________________ Date Submitted ______________________

____

____

Table 27-1
Cooling Rates for the Three Cases

Room Temperature (T_R) ____________ °C

Elapsed Time (min)	(s)	Uninsulated Case Temp(°C)	LnT	Air Gap Case Temp(°C)	LnT	Insulated Case Temp(°C)	LnT
0	0	______	______	______	______	______	______
5	300	______	______	______	______	______	______
10	600	______	______	______	______	______	______
15	900	______	______	______	______	______	______
20	1,200	______	______	______	______	______	______
25	1,500	______	______	______	______	______	______
30	1,800	______	______	______	______	______	______
35	2,100	______	______	______	______	______	______
40	2,400	______	______	______	______	______	______
45	2,700	______	______	______	______	______	______
50	3,000	______	______	______	______	______	______
55	3,300	______	______	______	______	______	______
60	3,600	______	______	______	______	______	______
65	3,900	______	______	______	______	______	______
70	4,200	______	______	______	______	______	______
75	4,500	______	______	______	______	______	______
80	4,800	______	______	______	______	______	______
85	5,100	______	______	______	______	______	______
90	5,400	______	______	______	______	______	______

Decay Constant (λ): ____________ ____________ ____________

Name ______________________________

For the uninsulated case, $(T_0 - T_R)$ = __________ and the equation for the temperature is:

____

For the air gap case, $(T_0 - T_R)$ = __________ and the equation for the temperature is:

____

For the insulated case, $(T_0 - T_R)$ = __________ and the equation for the temperature is:

____

EXPERIMENT 28

THE DIAMETER AND VOLUME OF A MOLECULE

Objective

To determine the approximate diameter and volume of the oleic acid molecule.

Equipment

Oleic acid solution (1 part oleic acid, 500 parts alcohol), large shallow tray, dropper, meter stick or ruler, lycopodium powder, chalk dust or cork dust.

Discussion and Theory

The size of a molecule is too small to measure directly with conventional measuring devices. We may, however, use conventional devices to determine the approximate size of a molecule indirectly.

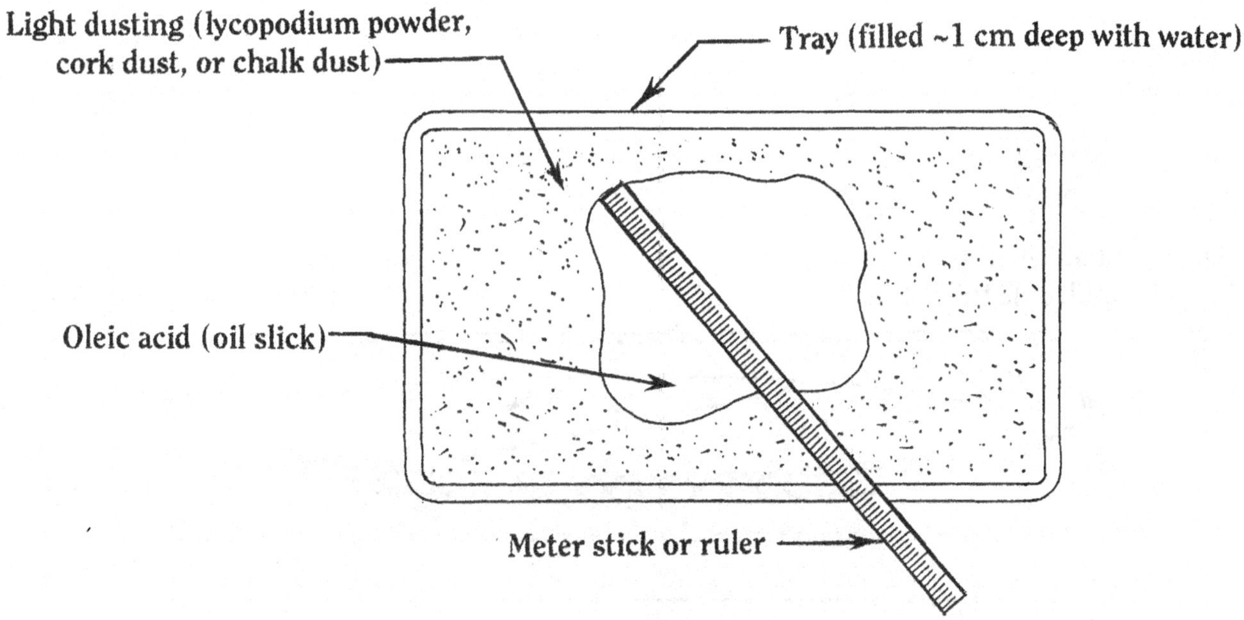

Figure 28-1 Measuring the Diameter of the Oil Slick of Oleic Acid

In this experiment, a large shallow tray is filled with water to a depth of about one centimeter as shown in Figure 28-1. Some lycopodium powder, chalk dust, or cork dust is lightly sprinkled on top of the water in the tray. A drop of a solution of one part oleic acid to 500 parts alcohol is gently dropped on the surface of the lightly powdered water. The alcohol goes into solution with the water leaving a one-molecule thick nearly circular oily film of oleic acid which displaces the powder. The diameter, d, of this oily film is measured with a meter stick or ruler.

The oleic acid film may be thought of as being a one-molecule thick disk on top of the water as shown in Figure 28-2. A disk with a diameter d and a thickness d_m (molecular diameter) is shown for the oil slick. The volume, V, of such a disk is

$$V = \frac{\pi d^2 d_m}{4} \qquad (28\text{-}1)$$

The area, A, of the top of the disk is

$$A = \frac{\pi d^2}{4} \qquad (28\text{-}2)$$

where d is the average diameter of the oleic acid film. Therefore, we may write equation (28-1) as

$$V = A\, d_m. \qquad (28\text{-}3)$$

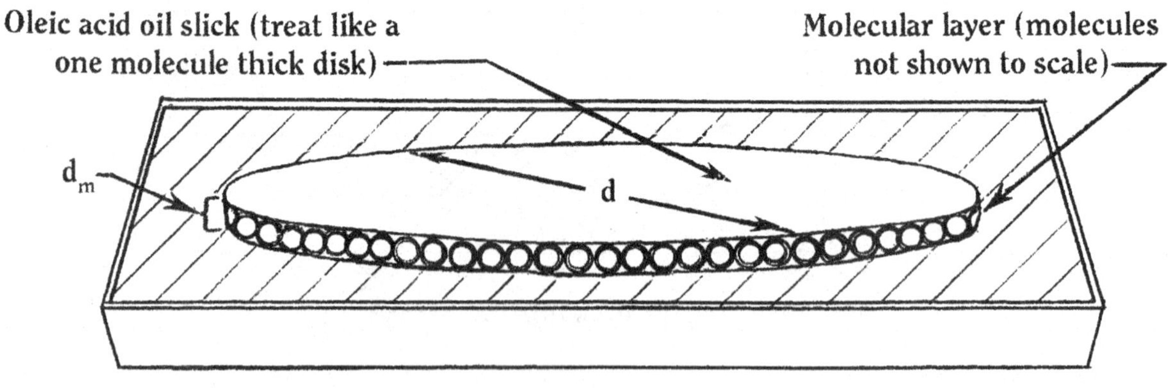

Figure 28-2 Treating the Oleic Acid Oil Slick as a One Molecule Thick Disk

If we now solve for the diameter of the molecule we obtain

$$d_m = V/A. \qquad (28\text{-}4)$$

From the diameter of the molecule we may find the volume of the oleic acid molecule. To do this, we first find the radius, $r_m = d_m/2$, of the molecule. We assume the volume, V_m, of the molecule is that of a sphere of radius r_m. Therefore,

$$V_m = (4/3)\ \pi\, r_m^{\ 3}. \qquad (28\text{-}5)$$

Procedure

1. Fill a tray with warm water to a depth of one centimeter.

2. Dust the surface lightly with either lycopodium powder, chalk dust or cork dust.

3. Carefully place one drop of the oleic acid solution in the center of the tray. You will see the oil drop spread into a thin film.

4. Take five readings of the diameter of the film. Determine the average diameter of the oil film. Record these values in Table 28-1.

5. Assume a circular surface area for the film. Calculate the area of this film from equation (28-2). Record your result in Table 28-1.

6. The volume of the oleic acid in the original drop is 5.0×10^{-5} cm^3. Divide this volume by the area calculated in step 5. This gives the thickness of the film which we assume is the diameter of the molecule, d_m. Record this value in Table 28-1.

7. Determine the radius of the oleic acid molecule. Record in Table 28-1.

8. Calculate the volume of an oleic acid molecule from the equation (28-5). Record in Table 28-1.

Questions

1. From your result for the size of the oleic acid molecule, how many molecules will have to be placed next to each other in a row to give you a length of 1.00 cm?

2. From the volume you obtained for the oleic acid molecule, how many molecules are there in one cubic centimeter?

3. The density is defined as the mass per unit volume (i.e., $D = M/V$). If the density of oleic acid is 0.89 g/cm^3, determine the mass of one molecule from the volume you obtained for the oleic acid molecule.

EXPERIMENT 29

ABSOLUTE ZERO

Objective

To determine absolute zero.

Equipment

A 250 ml flask with a stopper with a hole through it, a short glass tube that fits snugly through the stopper hole, a 600 ml beaker, a container large enough to hold enough water to fully submerge the 250 ml flask on its side, a Bunsen burner, a pole stand with a flask platform, burette holder, a Celsius thermometer, a graduated cylinder.

Discussion and Theory

Temperature cannot be decreased without limit. The coldest temperature possible is called *absolute zero*. Absolute zero is equal to -273 °C.

If the pressure is kept constant, changes in the temperature of a gas will cause its volume to change. If the volume of the gas is plotted as a function of its temperature, a straight line graph similar to Figure 29-1 is obtained. As the temperature is lowered the volume decreases. If the line is extended until the volume becomes zero, the value of absolute zero is obtained.

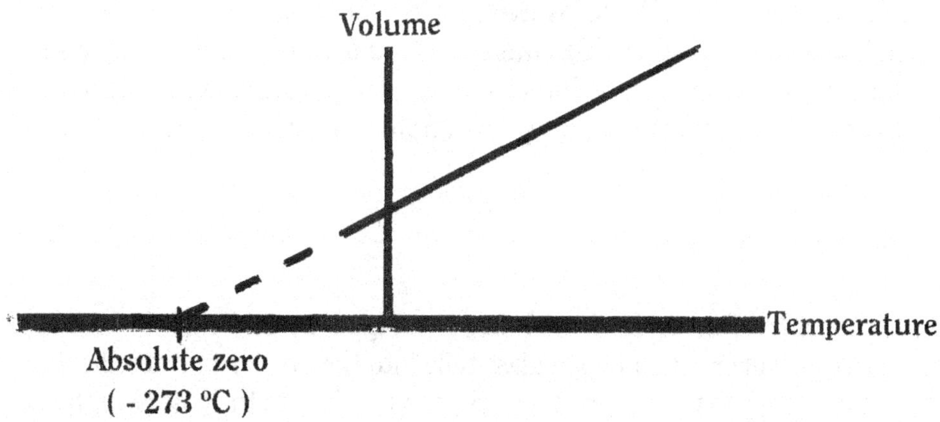

Figure 29-1 Volume of a Gas as a Function of the Temperature at Constant Pressure

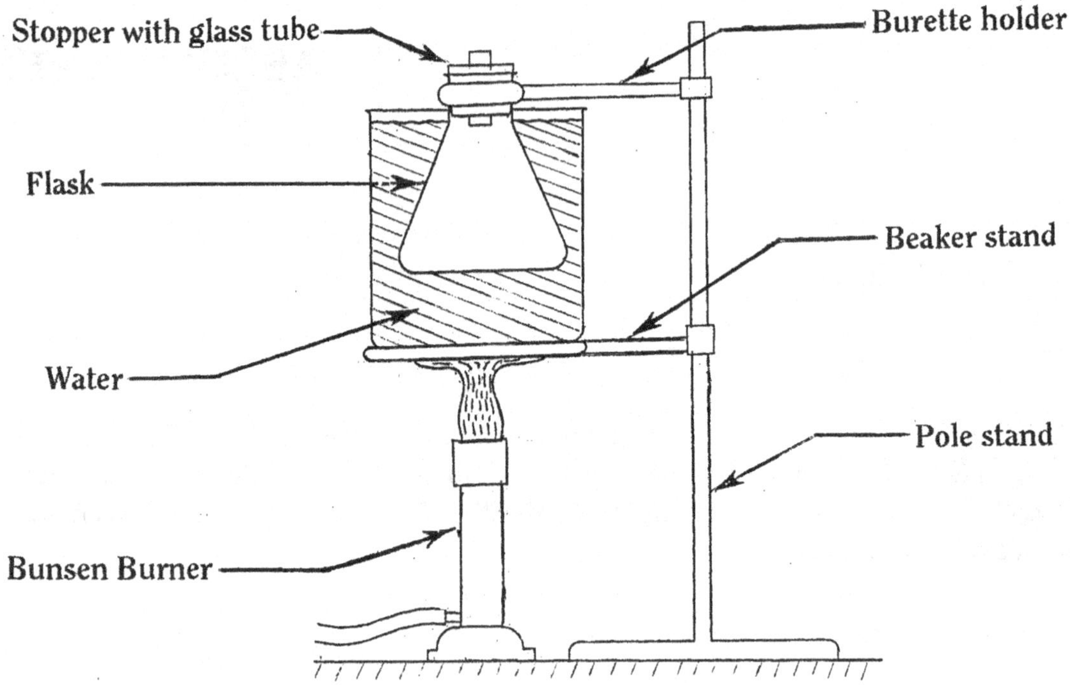

Figure 29-2 Heating the Flask with Boiling Water

In this experiment a dry flask is held submerged in water as shown in Figure 29-2. The water is heated and brought to a boil. After the air in the flask has had enough time to reach 100 °C, the flask is then submerged in cold water and left in the water until the air in the flask and the water are at the same temperature as shown in Figure 29-3. Water is free to enter the submerged flask through a small glass tube in the stopper. The amount of water that enters the flask is equal to the decrease in the volume of the air in the flask due to cooling. From the initial and final volumes and temperatures of the air in the flask it is possible to draw a graph similar to Figure 29-1, from which absolute zero is be determined.

Procedure

1. Set up the beaker stand and Bunsen burner as shown in Figure 29-2. Verify that the supplied 250 ml flask is totally dry. Place a stopper with an open glass tube into the dry flask. Place the flask inside a 600 ml beaker and place the beaker on the beaker stand. Secure the flask to the pole with the burette holder, as shown. Fill the beaker with enough water to cover the flask to within 2 to 3 centimeters of the burette holder.

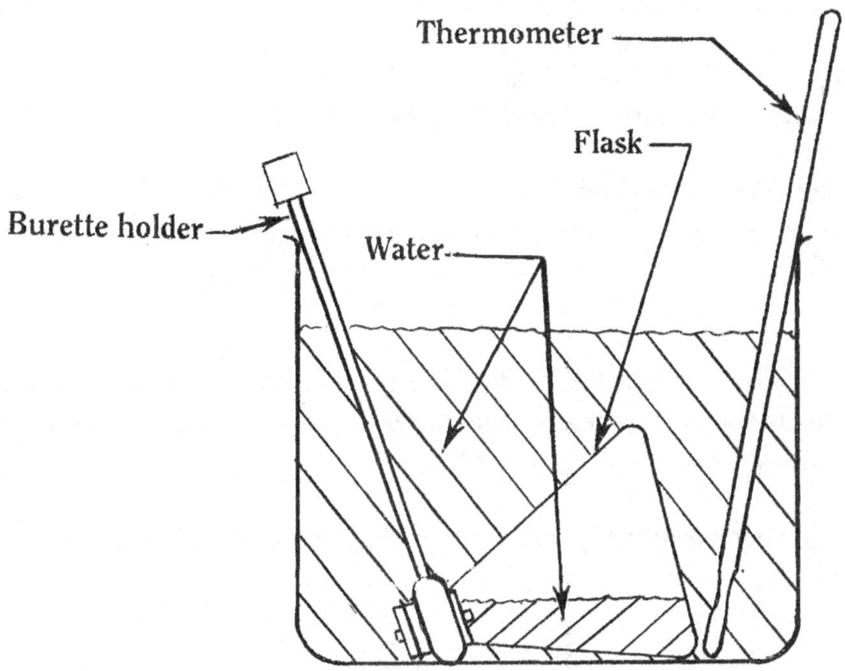

Figure 29-3 Cooling the Flask with Cold Water

2. Light the Bunsen burner. Heat the water in the beaker until it starts to boil. Let the water boil for at least 5 minutes, or so that the air within the flask attains a temperature of 100 °C.

3. Fill a container large enough to submerge the flask on its side with cold water.

4. Loosen the burette holder from the pole. Hold the burette holder with a towel and quickly remove the flask from the boiling water and place the flask into the cold water, glass tube first. Once in the water, fully submerge the flask on its side, as shown in Figure 29-3. Be certain that the top rim of the flask rests on its side on the bottom of the container so that the only water that enters the flask is due to the contraction of the air within the flask. A weight may be placed on top of the flask to keep it submerged on its side, or the flask may be held down with the burette holder. Place a thermometer in the water.

5. After at least 5 minutes have passed, read the temperature, T_2, of the water from the thermometer and remove the submerged flask from the water. Measure with a graduated cylinder the volume of water, V_w, that entered into the flask. Record T_2 and V_w in Table 29-1.

6. Empty the graduated cylinder. Fill the flask with water. Install the stopper into the flask; some water should overflow from the flask. Remove the stopper and measure the amount of water in the flask with the graduated cylinder. The total volume of this water is equal to the initial volume, V_1, of air in the flask. Record this volume as V_1 in Table 29-1.

7. Calculate the final volume of air in the cool flask, V_2, by subtracting the volume of water, V_w, that entered the cooled flask from the total volume of the flask, V_1, measured in step 6.

8. On a piece of graph paper construct a graph in which the data just obtained is plotted. The graph should have a volume scale on the y-axis that goes from zero up to at least V_1, and a temperature scale that goes from about -300 °C up to 100 °C on the x-axis. Plot the points $P_1(T_1,V_1)$ and $P_2(T_2,V_2)$ onto your graph.

9. Draw a straight line through points $P_1(T_1,V_1)$ and $P_2(T_2,V_2)$ that extends all the way to x-axis. Where this line intercepts the x-axis is your experimental value for absolute zero. Clearly indicate what this value is on your graph. Also record your result in Table 29-1.

10. Find the percent error in your experimental value for absolute zero. Recall that the standard value for absolute zero is - 273 °C.

Questions

1. Why was the stopper placed on the flask in step 6 of the procedure?

2. You were just asked to draw a straight line through points P_1 and P_2 on the graph in this experiment. From your experience in this and any other laboratory classes you may have had, if you had many more than just two points, how likely is it that all will fall exactly on a straight line? Explain why this is so.

3. If we could actually cool the air in the flask down to absolute zero, do you think the volume of the gas would actually go to zero? Would the air change in any way before actually reaching absolute zero?

4. You were told to verify that the flask was dry at the start of this experiment. Why?

5. If a flask with a larger volume were used in this experiment, how would the line obtained on the graph compare with the line you obtained in this experiment? Would you expect a larger flask to give the same value for absolute zero? Explain.

Name_______________________ Date Performed_______________

Lab. Partners__________________ Date Submitted_______________

____

____

Table 29-1
Determination of Absolute Zero

Initial Temperature of the air in the flask (T_1)	100 °C
Final Temperature of the air in the flask (T_2)	__________
Volume of water that entered flask (V_w)	__________
Initial volume of air in the flask (V_1)	__________
Final volume of air in the flask (V_2)	__________
Experimental value of absolute zero	__________
Percent error	__________

EXPERIMENT 30

METEOROLOGY - THE MOISTURE CONTENT OF AIR AND ATMOSPHERIC PRESSURE

Objectives

1. To learn how to convert the Celsius temperature into the Fahrenheit temperature.
2. To determine the dew point temperature.
3. To determine the relative humidity.
4. To determine the water content and vapor pressure of the air.
5. To measure atmospheric pressure.

Equipment

Sling psychrometer, Celsius thermometer, Fahrenheit thermometer, mercury barometer.

Discussion and Theory

The relationship between the Fahrenheit and Celsius temperature scales is

$$C = \frac{5}{9}(F - 32) \tag{30-1}$$

where C is the temperature in degrees Celsius and F is the temperature in degrees Fahrenheit.

In addition to the temperature meteorologists are concerned with the pressure of the atmosphere.

Barometric or Atmospheric Pressure is the pressure due to the gases of the atmosphere. It is measured by means of a barometer.

A *mercury barometer* consists of a glass tube, about 800 mm in length, that is open at one end and closed at the other as shown in Figure 30-1. This tube is completely filled with mercury and placed open end down into a cup filled with mercury. When this is done it is observed that the height of the mercury column drops to about 760 mm above the top of the mercury in the cup. The actual height of the mercury is used as a measure of atmospheric pressure.

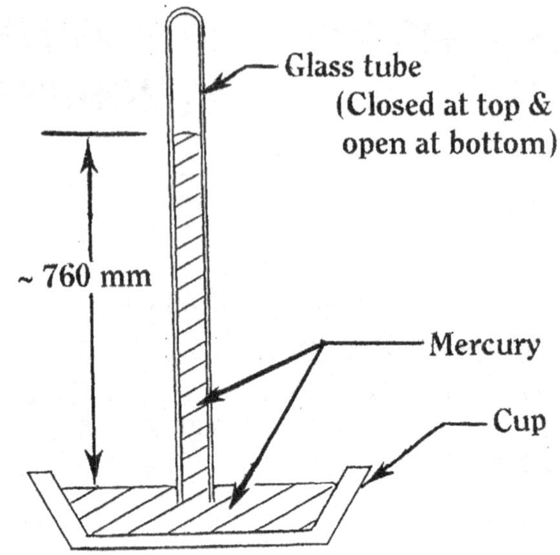

Figure 30-1 The Mercury Barometer

Standard atmospheric pressure at sea level is 760 mm of Mercury (Hg) or 29.92 inches of Mercury.

The actual reading of the barometric pressure may be higher or lower then standard atmospheric pressure. *High pressure* (high barometric pressure) is associated with fair weather. *Low pressure* (low barometric pressure) is associated with precipitation.

Meteorologists usually measure atmospheric pressure in terms of a unit called a *millibar* (mb). Standard atmospheric pressure at sea level is 1013 mb. The conversions from millimeters or inches of Mercury (Hg) are calculated as follows:

$$760 \text{ mm Hg} = 1013 \text{ mb}$$
$$1 \text{ mm Hg} = 1013 \text{ mb} / 760$$
$$1 \text{ mm Hg} = 1.333 \text{ mb.} \qquad (30\text{-}2)$$

Also,

$$29.92 \text{ inches Hg} = 1013 \text{ mb}$$
$$1 \text{ inch Hg} = 1013 \text{ mb} / 29.92$$
$$1 \text{ inch Hg} = 33.86 \text{ mb.} \qquad (30\text{-}3)$$

Example 1

At a particular time atmospheric pressure is measured to be 29.83 inches of Mercury. To convert this to millibars multiply by 33.86 mb/(inch Hg).

(29.83 inch Hg) X (33.86 mb / inch Hg) = 1010 mb.

To convert this to mm Hg multiply by (1 mm Hg)/(1.333 mb).

(1010 mb) X [(1 mm Hg) / (1.333 mb)] = 758 mm Hg.

In addition to the temperature and the pressure, weather also depends on the amount of moisture in the air. To study the effect moisture has on weather several terms need to be understood.

Saturated air is air which contains the maximum amount of moisture possible at a particular temperature. The maximum amount of moisture that air can hold increases with the temperature as can be seen in Table 30-1 at the end of this experiment..

Example 2

From Table 30-1 we see that at 10 °C (first column) the maximum amount of moisture that air can hold is 9.40 grams per cubic meter (second column). At 20 °C the maximum is 17.30 grams per cubic meter. Notice that as the temperature increases the amount of moisture that air is able to hold also increases.

The vapor pressure is that part of atmospheric pressure due to the water vapor present in the air

The saturated vapor pressure is the vapor pressure of saturated air. The saturated vapor pressure increases with the temperature as can be seen in Table 30-1).

Example 3

From Table 30-1 we see that at 10 °C (first column) the pressure due to water vapor in saturated air, the saturated vapor pressure, is 12.27 mb (third column). At 20 °C the saturated vapor pressure is 23.37 mb. Notice that as the temperature increases the saturated vapor pressure also increases.

The dew point temperature (or just *the dew point*) is the temperature to which the air would have to be cooled to become saturated air.

The relative humidity is the actual amount of moisture present in the air divided by the maximum amount of moisture that the air can hold. The relative humidity is usually written as a percent.

The relative humidity may be expressed in either of two ways. It may expressed as the actual amount of water present per unit volume, W_A, divided by the water that would be present per unit volume if the air were saturated, W_S or it may be expressed as the actual vapor pressure, V_A divided by the saturated vapor pressure, V_S; the quantities W_A and W_S are given in g/m^3; V_A and V_S are given in mb.

Therefore, the relative humidity, R.H., may be calculated from

$$R.H. = \frac{W_A}{W_S} X\ 100, \tag{30-4}$$

or from

$$R.H. = \frac{V_A}{V_S} X\ 100. \tag{30-5}$$

From the relative humidity, R.H., it is possible to determine the actual amount of water in grams per cubic meter, from the number of grams per cubic meter of water in saturated air in Table 30-1 by solving for W_A in equation 30-4 to obtain,

$$W_A = \frac{(R.H.)\ X\ W_S}{100}. \tag{30-6}$$

The actual vapor pressure, V_A, may be determined from the saturated vapor pressure, V_S, in Table 30-1 by solving for V_A in equation 30-5 to obtain

$$V_A = \frac{(R.H.)\ X\ V_S}{100}. \tag{30-7}$$

From the actual amount of water per cubic meter, or from the actual vapor pressure, the dew point temperature may be determined by finding the temperature to which the air would have to be cooled to become saturated.

Example 4

(a) Assume that the temperature is 23 °C and the relative humidity is 62 %. From Table 30-1 we see that at 23 °C saturated air would have 20.58 g of water per cubic meter of air. Since the relative humidity is 62 %, the actual amount of water per cubic meter is calculated from equation (30-6) to be

$$W_A = (R.H.)\ X\ W_S / 100 = 62\ X\ 20.58 / 100 = 12.76\ g/m^3.$$

(b) To find the dew point from the actual amount of water per cubic meter, notice in Table 30-1 that at 15 °C the amount of moisture in saturated air is 12.83 g/m^3 which is close to the 12.76 g/m^3 actually in the air. Therefore, if the air is cooled from 23 °C to about15 °C the air will become saturated. The dew point is therefore about 15 °C.

(c) From Table 30-1 we see that at 23 °C the saturated vapor pressure is 28.13 mb. The actual vapor pressure is calculated from equation (30-7) to be

$$V_A = (R.H.) \times V_S / 100 = 62 \times 28.13 / 100 = 17.44 \text{ mb}.$$

(d) To find the dew point temperature from the vapor pressure notice that in Table 30-1 at 15 °C the vapor pressure is 17.07 mb which is close to 17.44 mb. Therefore, if the air is cooled from 23 °C to about 15 °C the air will become saturated. The dew point is therefore about 15 °C.

In this experiment we will use a device called the *sling psychrometer*. It consists of two parallel thermometers which are mounted on a platform. One of the thermometers, called the *wet bulb thermometer*, has a wick over its bulb which is kept wet when the device is used. The other thermometer, called the *dry bulb thermometer*, has a bare bulb which is kept dry when the device is used. The sling psychrometer is attached to a shaft and should be rotated in a circle about this shaft for one or two minutes. The *dry bulb temperature* is then read from the dry bulb thermometer and the *wet bulb temperature* is read from the wet bulb thermometer.

It is possible to determine the relative humidity and the dew point from the wet bulb temperature, T_W, and dry bulb temperature, T_D. To do this, the wet bulb depression, ΔT, needs to be calculated

$$\Delta T = T_D - T_W \tag{30-8}$$

From the dry bulb temperature (which, by the way, is the actual temperature) and the wet bulb depression the relative humidity may be determined from Table 30-2 and the dew point may then be determined from Table 30-3. Both tables may be found at the end of this experiment.

Example 5

A dry bulb temperature of 23 °C and a wet bulb temperature of 18 °C are obtained from a sling psychrometer. The wet bulb depression is

$$\Delta T = T_D - T_W = 23 - 18 = 5 \text{ °C}.$$

(a) To find the relative humidity from Table 30-2 read down the "Dry Bulb Temp" column (column 1) to find 23 °C and go across to the column for a 5 °C Wet Bulb Depression (column 6) and read 62. The relative humidity is 62%.

(b) To find the dew point temperature from Table 30-3 read down the "Dry Bulb Temp" column (column 1) to find 23 °C and go across to the column for a 5 °C Wet Bulb Depression (column 6) and read 15. The dew point temperature is 15 °C.

Procedure

The following procedure is to be performed for readings taken inside your laboratory and outdoors:

1. Measure the temperature with a Celsius and a Fahrenheit thermometer and record these temperatures in Table 30-4. Use equation (30-1) to convert the Fahrenheit temperature to degrees Celsius and record. Is this the same as the temperature read from the Celsius thermometer?

2. Wet the wick of the sling psychrometer. Be careful not to wet the bulb of the other thermometer. Hold the sling psychrometer firmly by its handle and rotate it vigorously for a minute or two. Be careful not to hit anything. Read the wet and dry bulb temperatures, T_W and T_D, and record in Table 30-5.

3. Determine the wet bulb depression, ΔT, from equation (30-8) and record in Table 30-5.

4. From Table 30-2 determine the relative humidity and record in Table 30-5. Column 1 lists the dry bulb temperature. The next 22 columns list the relative humidity corresponding to the wet bulb depressions indicated across the top of the columns. See *Example 5(a).*

5. For your dry bulb temperature (the actual temperature), determine the number of grams per cubic meter of water in saturated air, W_S, from Table 30-1 and record in Table 30-5. See *Example 2.*

6. For your dry bulb temperature, determine the vapor pressure of saturated air, V_S, from Table 30-1 and record in Table 30-5. See *Example 3.*

7. From the relative humidity determined in step 4, calculate the actual amount of water per cubic meter of air and the actual vapor pressure. Use equations (30-6) and (30-7). Record in Table 30-5. See *Example 4(a)* and (c).

8. From the dry bulb temperature and the wet bulb depression, determine the dew point temperature. Table 30-3 lists the dew point temperatures under the wet bulb depressions for each dry bulb temperature shown in column 1. Record your result in Table 30-5. See *Example 5(b).*

9. From your calculated value for the actual amount of moisture, W_A, in the air determine the dew point temperature from Table 30-1. The dew point will be that temperature at which W_S will make the air saturated. See *Example 4(b).*

10. Determine the barometric pressure from a barometer in inches of mercury and in millimeters of mercury. Record in Table 30-6. Convert each of these values to millibars and record in Table 30-6. See *Example 1.*

Questions

1. You calculated the inside and outside temperatures in degrees Celsius from the temperatures in degrees Fahrenheit. Did these results agree with your measured values (within experimental error)?

2. You determined the dew point temperatures two ways in this experiment. Were the results the same (within experimental error)?

3. You calculated atmospheric pressure in millibars from your readings in millimeters of mercury and inches of mercury. If both readings were taken from the same barometer, how large a discrepancy would you expect between your calculated values in millibars (Would you expect relatively small or large discrepancies). Were your results the same (within experimental error)?

4. Barometric pressure was only determined indoors. How does this value compare with the barometric pressure outdoors?

Table 30-1
Moisture Content of Air

Temp. (°C)	Water per $meter^3$ of Air (g/m^3)	Vapor Pressure (mb)	Temp. (°C)	Water per $meter^3$ of Air (g/m^3)	Vapor Pressure (mb)
-10	2.36	2.86	16	13.63	18.17
-9	2.54	3.10	17	14.84	19.40
-8	2.74	3.35	18	15.37	20.63
-7	2.95	3.63	19	16.21	22.00
-6	3.17	3.91	20	17.30	23.37
-5	3.41	4.23	21	18.34	24.90
-4	3.66	4.55	22	19.43	26.43
-3	3.93	4.91	23	20.58	28.13
-2	4.22	5.28	24	21.78	29.83
-1	4.52	5.68	25	23.05	31.72
0	4.85	6.11	26	24.38	33.61
1	5.19	6.58	27	25.78	35.70
2	5.56	7.05	28	27.24	37.80
3	5.95	7.59	29	28.78	40.11
4	6.36	8.13	30	30.38	42.43
5	6.80	8.74	31	32.07	44.99
6	7.26	9.35	32	33.83	47.55
7	7.75	10.03	33	35.68	50.37
8	8.27	10.72	34	37.61	53.20
9	8.82	11.49	35	39.63	56.31
10	9.40	12.27	36	41.75	59.42
11	10.01	13.14	37	43.96	62.84
12	10.66	14.02	38	46.26	66.26
13	11.35	15.00	39	48.67	70.02
14	12.07	15.98	40	51.20	73.78
15	12.83	17.07			

Table 30-2
The Relative Humidity from the Sling Psychrometer

	Wet Bulb Depression																					
Dry Bulb Temp.	1	2	3	4	5	6	7	8	9	10	11	12	13	14	15	16	17	18	19	20	21	22
−10°C	67	35																				
−9	69	39	9																			
−8	71	43	15																			
−7	73	46	20																			
−6	74	49	25																			
−5	76	52	29	11																		
−4	77	55	33	12																		
−3	78	57	37	17																		
−2	79	60	40	22																		
−1	81	62	43	26	8																	
0	81	64	46	29	13																	
1	83	66	49	33	17																	
2	84	68	52	37	22	7																
3	84	70	55	40	26	12																
4	85	71	57	43	29	16																
5	86	72	58	45	33	20	7															
6	86	73	60	48	35	24	11															
7	87	74	62	50	38	26	15															
8	87	75	63	51	40	29	19	8														
9	88	76	64	53	42	32	22	12														
10	88	77	66	55	44	34	24	15	6													
11	89	78	67	56	46	36	27	18	9													
12	89	78	68	58	48	39	29	21	12													
13	89	79	69	59	50	41	32	23	15	7												
14	90	79	70	60	51	42	34	26	18	10												
15	90	80	71	61	53	44	36	27	20	13	6											
16	90	81	71	63	54	46	38	30	23	15	8											
17	90	81	72	64	55	47	40	32	25	18	11											
18	91	82	73	65	57	49	41	34	27	20	14	7										
19	91	82	74	65	58	50	43	36	29	22	16	10										
20	91	83	74	66	59	51	44	37	31	24	18	12	6									
21	91	83	75	67	60	53	46	39	32	26	20	14	9									
22	92	83	76	68	61	54	47	40	34	28	22	17	11	6								
23	92	84	76	69	62	55	48	42	36	30	24	19	13	8								
24	92	84	77	69	62	56	49	43	37	31	26	20	15	10	5							
25	92	84	77	70	63	57	50	44	39	33	28	22	17	12	8							
26	92	85	78	71	64	58	51	46	40	34	29	24	19	14	10	5						
27	92	85	78	71	65	58	52	47	41	36	31	26	21	15	12	7						
28	93	85	78	72	65	59	53	48	42	37	32	27	22	18	13	9	5					
29	93	86	79	72	66	60	54	49	43	38	33	28	24	19	15	11	7					
30	93	86	79	73	67	61	55	50	44	39	35	30	25	21	17	13	9	5				
32	93	86	79	74	68	62	57	51	46	41	37	32	28	24	20	16	12	9	5			
34	93	87	81	75	69	63	58	53	48	43	39	35	30	26	23	19	15	12	8	5		
37	94	87	82	76	70	65	60	55	51	46	42	38	34	30	26	23	19	16	13	10	5	
40	94	88	82	77	72	67	62	57	53	48	44	40	36	33	29	26	23	20	16	14	10	5

Table 30-3
The Dew Point from the Sling Psychrometer

Dry Bulb Temp. °C	Wet Bulb Depression 1	2	3	4	5	6	7	8	9	10	11	12	13	14	15	16	17	18	19	20	21	22
–10	–14	–22																				
– 9	–13	–20																				
– 8	–12	–18	–29																			
– 7	–11	–16	–25																			
– 6	–10	–14	–22																			
– 5	– 9	–13	–20																			
– 4	– 7	–11	–17	–29																		
– 3	– 6	–10	–15	–25																		
– 2	– 5	– 8	–13	–20																		
– 1	– 4	– 7	–11	–17																		
– 0	– 3	– 6	– 9	–15	–24																	
1	– 2	– 5	– 8	–13	–20																	
2	– 1	– 3	– 6	–11	–17																	
3	0	– 2	– 5	– 9	–14																	
4	1	– 1	– 4	– 7	–11	–19																
5	3	0	– 3	– 5	– 9	–15																
6	4	1	– 1	– 4	– 7	–13	–21															
7	5	2	0	– 3	– 6	–11	–18															
8	6	3	1	– 2	– 5	– 9	–14															
9	7	4	3	0	– 3	– 7	–11															
10	8	6	4	1	– 2	– 5	– 9	–14	–28													
11	9	7	5	2	0	– 4	– 7	–11	–22													
12	10	8	6	4	1	– 2	– 5	– 9	–16													
13	11	9	7	5	2	0	– 4	– 7	–13													
14	12	11	9	6	4	1	– 2	– 5	–10	–17												
15	13	12	10	7	5	2	0	– 3	– 8	–13												
16	14	13	11	9	7	4	1	– 1	– 6	–10	–17											
17	15	14	12	10	8	5	2	0	– 4	– 7	–14											
18	16	15	13	11	9	7	4	2	– 2	– 5	–10	–19										
19	17	16	14	12	10	8	5	3	0	– 3	– 7	–15										
20	19	17	15	14	12	10	7	4	2	– 2	– 5	–10	–19									
21	20	18	16	15	13	11	8	6	3	0	– 3	– 7	–15									
22	21	19	17	16	14	12	10	8	5	3	– 1	– 5	–10	–19								
23	22	20	18	17	15	13	11	9	6	4	0	– 3	– 7	–15								
24	23	21	20	18	16	14	12	10	8	6	2	– 1	– 5	–10	–18							
25	24	22	21	19	17	15	13	11	9	7	5	1	– 2	– 7	–14							
26	25	23	22	20	18	17	15	13	11	9	6	3	0	– 4	– 9	–18						
27	26	24	23	21	19	18	16	14	13	10	7	5	2	– 2	– 6	–13						
28	27	25	24	22	21	19	17	16	14	11	9	7	4	1	– 3	–9	–18					
29	28	26	25	23	22	20	18	17	15	12	10	8	6	3	– 1	– 3	–13					
30	29	27	26	24	23	21	19	18	16	14	12	10	8	5	1	– 2	– 8	–15				
31	30	28	27	25	24	22	20	19	17	15	13	11	9	6	3	0	– 6	–11				
32	31	29	28	27	25	24	22	21	19	17	15	13	11	8	5	2	– 2	– 7	–14			
33	32	30	29	28	26	25	23	22	20	18	16	14	12	10	7	4	0	– 4	–10			
34	33	31	30	29	27	26	24	23	21	20	18	16	14	12	9	6	3	1	– 5	–12	–29	
35	34	32	31	30	28	27	25	24	22	21	19	17	15	13	11	8	5	2	– 3	– 8	–20	
36	35	33	32	31	29	28	27	25	24	22	20	19	17	15	13	10	7	4	0	– 4	–10	
37	36	34	33	32	30	29	28	26	25	23	21	20	18	16	14	11	9	6	3	– 3	– 7	
38	37	35	34	33	32	30	29	28	26	25	23	21	19	17	15	13	11	8	5	– 1	– 3	– 9
39	38	36	35	34	33	31	30	29	27	26	24	22	20	18	16	14	12	10	6	2	0	– 3
40	39	37	36	35	34	32	31	30	28	27	25	24	22	20	18	16	14	12	8	6	2	– 2

Name ____________________	Date Performed ____________________
Lab. Partners ____________________	Date Submitted ____________________
____________________	
____________________	

Table 30-4
Indoor and Outdoor Temperatures in °C and °F

	°C	°F	°C (Calculated from °F)
Room Temperature	________	________	________
Outside Temperature	________	________	________

Table 30-5
The Moisture in the Air and the Dew Point

	Inside	Outside
Wet Bulb Temperature (T_W)	________	________
Dry Bulb Temperature (T_D)	________	________
Wet Bulb Depression (ΔT)	________	________
Relative Humidity (R.H.)	________	________
Water in Saturated Air (W_S)	________	________
Saturated Vapor Pressure (V_S)	________	________
Actual Water in Air (W_A)	________	________
Actual Water Vapor in Air (V_A)	________	________
Dew Point (From Table 30-3)	________	________
Dew Point (From Table 30-1)	________	________

Table 30-6
Barometric Pressure

Barometric Pressure	________mm	________mb
Barometric Pressure	________inches	________mb

EXPERIMENT 31

REFLECTION

Objective

To study the law of reflection

Equipment

Graph paper, board, mirror, 2 mirror supports, 3 pins, a protractor, and a ruler.

Discussion and Theory

A ray of light that falls on a reflecting surface is called an *incident ray*, as shown in Figure 31-1. The angle the incident ray makes with the normal to the reflecting surface (i.e., the perpendicular to the reflecting surface) is called the *angle of incidence*. The ray that comes off the surface is called the *reflected ray*. The angle the reflected ray makes with the normal to the reflecting surface is called the *angle of reflection*.

A fundamental law of optics is the *law of reflection*. This law states that *for an incoming ray of light the angle of incidence is equal to the angle of reflection*. The incident ray, the reflected ray and the normal are all in the same plane.

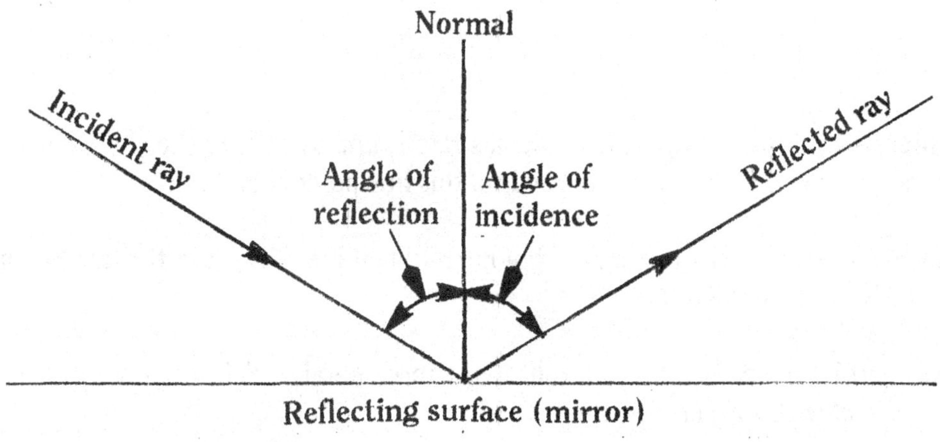

Figure 31-1 A Ray Reflected from Reflecting Surface

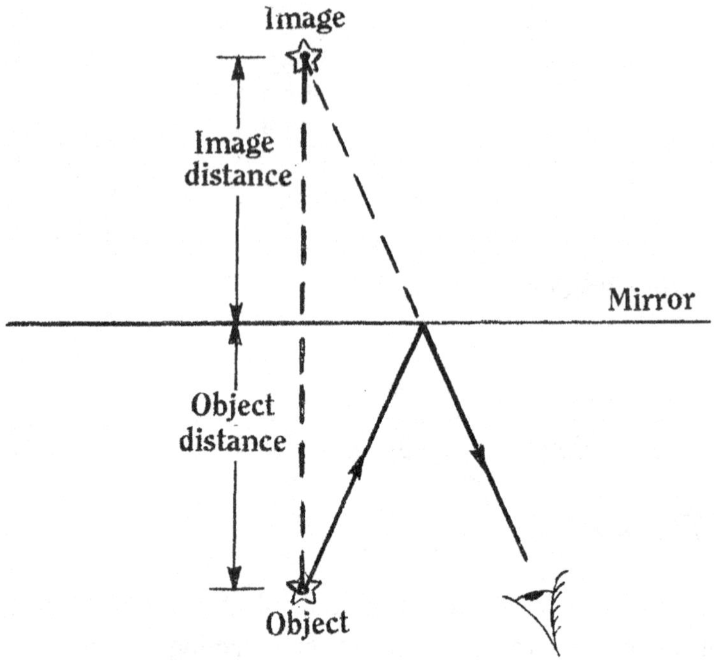

Figure 31-2 The Object, Image, Object Distance, and Image Distance

If an *object* is placed in front of a mirror an *image* of the object can be seen in the mirror as shown in Figure 31-2. The perpendicular distance from the object to the mirror is called the *object distance.* The perpendicular distance from the apparent location of the image to the mirror is called the *image distance.* For a plane mirror the image distance is equal to the object distance.

Procedure

1. In the center of a sheet of graph paper, draw a straight line parallel to the shorter edges of the paper along one of the graph paper lines. Label the ends of this line point M and M'.

2. Place a silvered mirror so that the back of the mirror is on line MM'. Use the mirror supports to hold the mirror as shown in Figure 31-3.

3. Center a pin (pin 1) about 7 or 8 cm in front of the mirror and stick it into the paper. Label this point O. Pin 1 is the object in this experiment.

4. Stick a second pin (pin 2) into the paper several centimeters to the right of pin 1.

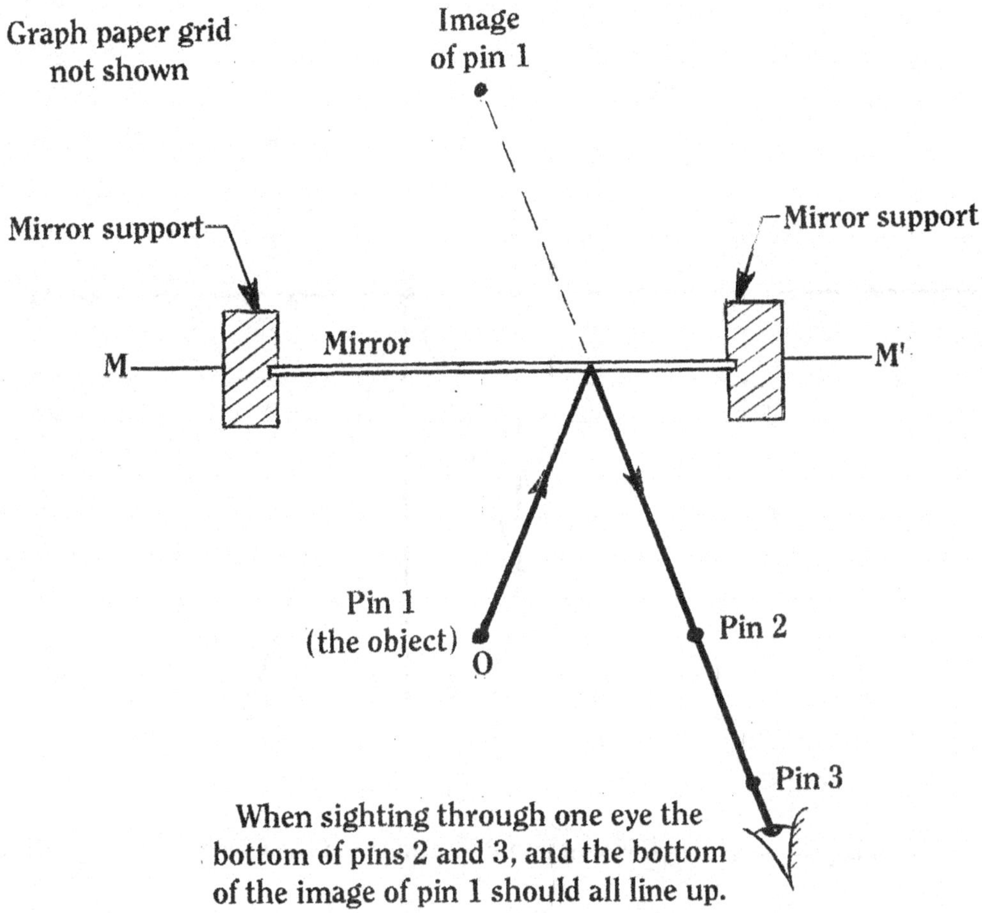

Figure 31-3 Performing the Experiment

5. Sight through one eye along the bottom of pin 2 and the bottom of the image of pin 1 seen in the mirror. Place a third pin (pin 3) along this line of sight so that the bottoms of pins 2 and 3 and the bottom of the image of pin 1 are all in a straight line as shown in Figure 31-3.

6. Remove the pins and the mirror from the paper. Label the pin holes made by pins 2 and 3 as points 2 and 3.

7. Draw a line from point 3 through point 2 until it meets line MM'. Label the point of intersection A as shown in Figure 31-4.

8. Draw a line from point O to point A as shown in Figure 31-4. When you lined-up the image of pin 1 with pins 2 and 3, a ray from point O (the incident ray) was reflected at point A past points 2 and 3 (the reflected ray) and into your eye.

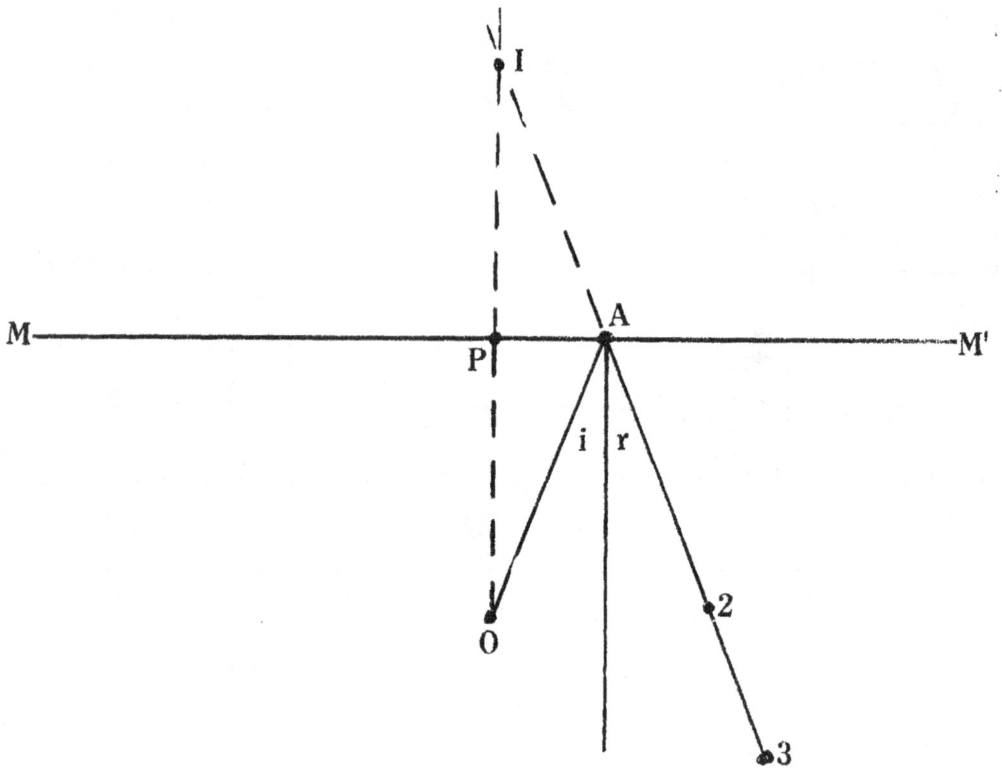

Figure 31-4 Finding the Angles of Incident and Reflection, and the Object and Image Distances

9. At point A, carefully draw a perpendicular from line MM' that divides angle OA2 into two angles, as shown in Figure 31-4. The angle between the incident ray OA and the perpendicular is the angle of incidence, i. The angle between the reflected ray A2 and the perpendicular is the angle of reflection, r. Measure the angles of incidence and reflection and record them on the graph paper and in Table 31-1

10. Theoretically, the angles of incidence and reflection should be the same. Let the angle of incidence be the standard value for the angle of reflection; determine the percent error in the angle of reflection and record in Table 31-1.

11. Drop a perpendicular from point O to line MM' (dashed line), as shown in Figure 31-4. Label the intersection of this perpendicular to line MM' as point P. Extend line OP behind the mirror.

12. Extend the line through points 2, 3, and A until it (dashed line) crosses the extension of line OP drawn in step 11. Label the intersection of these lines point I. Point I is the location of the image of the object at point O (pin 1).

13. Measure the image distance, PI, and the object distance, PO. Record your results in Table 31-1.

14. The image distance and the object distance are theoretically equal. Let the object distance be the standard value for the image distance; find the percent error in the image distance and record in Table 31-1.

15. Place your graph paper back onto the board. Place the mirror over line MM' as before. Re-insert pin 1 in the pinhole at point O. Stick a second pin (pin 2') into the paper several centimeters to the left of pin 1. Repeat steps 5 to 14 on the left side of pin 1. Label all points as before, but with primes (e.g., points 2,3 and A become points 2', 3' and A' on the left side).

Questions

1. When you did the experiment on the right and left sides, did the image distances come out to be exactly the same? Should they have come out to be the same?

2. If pin 1 was kept at the same point O that you used, but you placed pin 2 further to the side, then how would your value for the angle of incidence change? How would your value for the angle of reflection change?

3 If you followed the procedure spelled out in this experiment exactly as stated except that your mirror was slightly inclined to the line MM', how would your results be affected?

4. If the normal to point A in Figure 31-4 is not properly drawn and is not actually perpendicular, then what effect would it have on your results for the law of reflection? Would it affect the result that the image distance should equal the object distance?

5. If the surface of the mirror were curved like a dish, would the image distance still equal the object distance? (Hint: When you look into a curved mirror what do you see?)

Name______________________________ Date Performed______________________
Lab Partners ______________________ Date Submitted______________________
____
____

Table 31-1
Reflection from a Plane Surface

Right Side

Angle of Incidence (i)	________	Angle of Reflection (r)	________	% error	________
Object distance (PO)	________	Image Distance (PI)	________	% error	________

Left Side

Angle of Incidence (i')	________	Angle of Reflection (r')	________	% error	________
Object distance (P'O')	________	Image Distance (P'I')	________	% error	________

EXPERIMENT 32

REFRACTION

Objective

To study the refraction of light.

Equipment

Refraction apparatus

Discussion and Theory

When light passes from one medium to another its speed changes. The *index of refraction,* n, is defined as

$$n = \frac{\text{speed of light in vacuum}}{\text{speed of light in the medium}} \qquad (32\text{-}1)$$

The speed of light in vacuum is 3×10^8 m/s. The indices of refraction of some common substances are listed in Table 32-1

Table 32-1
The Indices of Refraction of Various Substances

Air	1.0003
Crown glass	1.52
Diamond	2.42
Ice	1.31
Water	1.33

When a ray of light passes from one substance into another substance, it is observed to bend. This bending of a ray of light is called *refraction.* The *Law of Refraction,* also known as *Snell's Law,* says that when a ray of light passes through the interface between two substances then (See Figure 32-1)

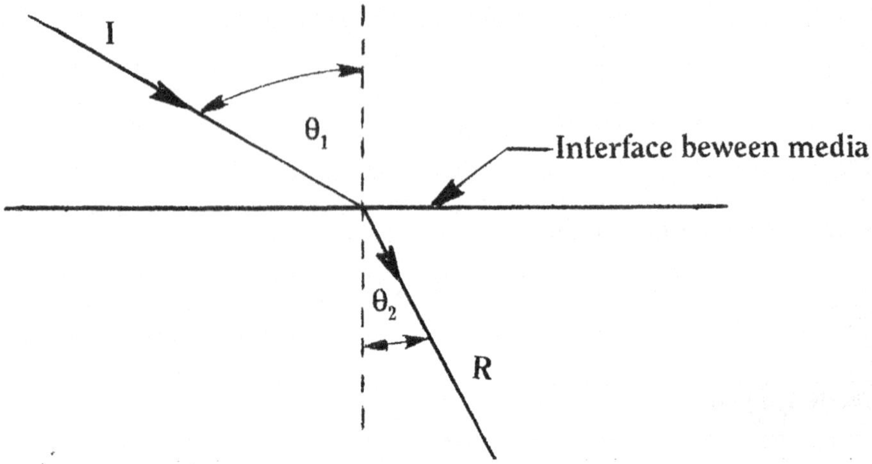

Figure 32-1 The Refraction of a Ray of Light

$$n_1 \sin \theta_1 = n_2 \sin \theta_2 \qquad (32\text{-}2)$$

where

n_1 = the index of refraction in medium 1.

n_2 = the index of refraction in medium 2.

θ_1 = the angle the ray of light makes in medium 1 with the perpendicular to the interface between the two media.

θ_2 = the angle the ray of light makes in medium 2 with the perpendicular to the interface between the two media.

The ray of light that strikes the interface is called the *incident ray*. The angle this ray makes with the normal to the interface is called the *angle of incidence.* In Figure 32-1, I is the incident ray and θ_1, is the angle of incidence.

The ray of light that emerges from the interface into the second medium is called the *refracted ray*. The angle the refracted ray makes with the normal to the interface is called the *angle of refraction.* In Figure 32-1, R is the refracted ray and θ_2, is the angle of refraction.

In Figure 32-1, the light ray is incident from the top. The law of refraction is just as valid for a ray incident from the bottom. If the ray in Figure 32-1 is reversed it will travel along the exact same path as the ray shown , but in the opposite direction.

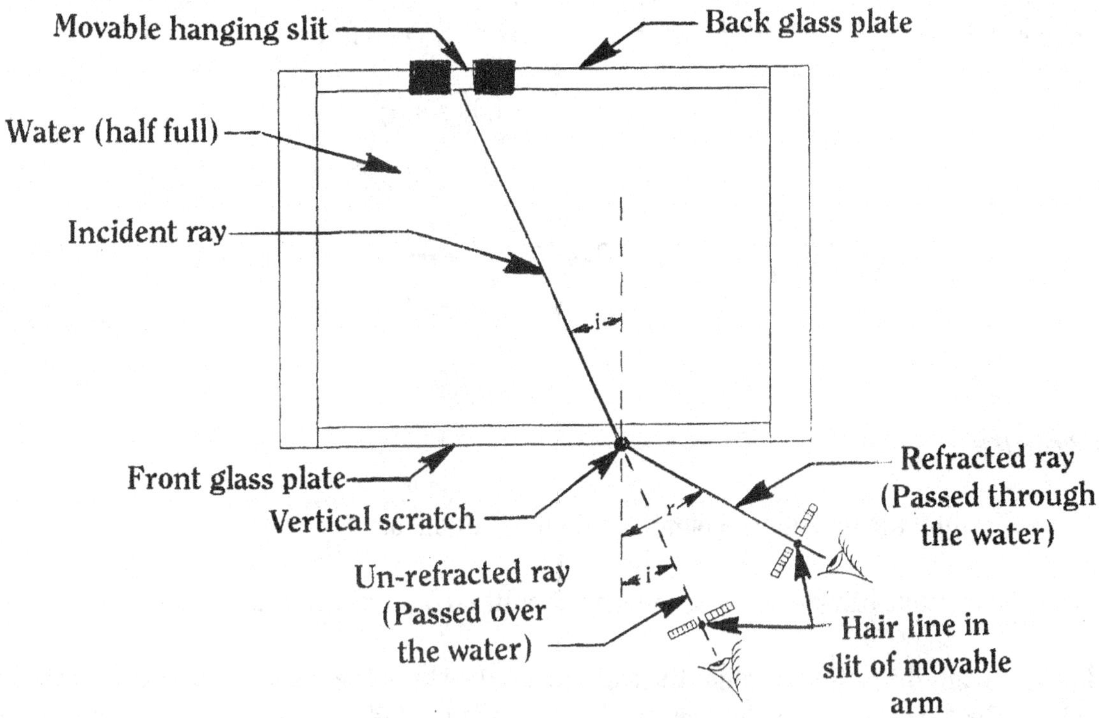

Figure 32-2 Using the Apparatus to Determine the Angle of Incidence and the Angle of Refraction

In this experiment we will use the refraction apparatus to determine the index of refraction of water. The front and rear plates of the tank of the apparatus are made of glass as shown in Figure 32-2. A movable vertical slit hangs on the rear glass plate. A vertical scratch exists on the front plate. A slit on the moveable arm is used for sighting.

To use the apparatus, the tank is half filled with water. A sighting is made by looking through the water through the slit on the movable index arm and aligning it with the vertical scratch on the front plate and the movable slit on the rear plate of the apparatus, as shown in Figure 32-2. The ray of light that emerges from the tank and enters the moveable index arm is refracted as it leaves the tank. The angle the refracted ray makes with the normal to the vertical scratch is measured from the angle scale under the movable arm. This angle is recorded as the angle of refraction, r.

A second sighting is made by looking above the water through the slit on the movable arm and aligning it with the vertical scratch on the front plate and the movable slit on the rear plate of the apparatus. The ray of light that emerges from the tank and enters the movable index arm travels in a straight line (it is not refracted). The angle this ray makes with the normal to the vertical scratch is measured from the angle scale under the index arm. This angle is recorded as the angle of incidence, i.

If we apply equation (32-2) to the refraction apparatus and let medium 1 be water, and medium 2 be air, then $\theta_1 = i$, the angle of incidence; $n_1 = n_{water}$, the index of refraction of water; $\theta_2 = r$, the angle of refraction; and $n_2 = n_{air} = 1.00$, the index of refraction of air. Therefore,

$$n_{water} \sin (i) = 1.00 \sin (r)$$

and

$$n_{water} = \frac{\sin (r)}{\sin (i)} . \qquad (32\text{-}3)$$

Procedure

1. Pour water into the refraction apparatus until it is half filled.

2. Place the movable hanging slit to a convenient position on the back plate of the apparatus.

3. Look through the water and align the hair line on the slit of the movable index arm with the vertical scratch on the front plate and the movable hanging slit located on the back plate of the apparatus, as shown in Figure 32-2.

4. Read the angle of the movable arm. This is the angle of refraction, r. Record in Table 32-2.

5. Look above the water and align the hair line on the slit of the movable index arm with the vertical scratch on the front plate and the movable hanging slit on the back of the apparatus, as shown in Figure 32-2. This is the angle of incidence , i. Record in Table 32-2.

6. Determine sin (i) and sin (r), and record in Table 32-2.

7. Calculate the index of refraction from equation (32-3) and record in Table 32-2.

8. Determine your percent error and record in Table 32-2. The standard value of the index of refraction for water is 1.33.

9. Repeat steps 2 through 8 for two other positions for the movable slit located on the back plate.

Questions

1. Consider the results from the first run of this experiment. If the ray of light is reversed so that it is incident on the apparatus with an angle of incidence equal to your angle of refraction, what will be the angle of refraction of the reversed ray.

2. What is the speed of light in water?

EXPERIMENT 33

LENSES

Objectives

1. To find the focal point for each of several converging lenses.
2. To learn and apply the lens equation.
3. To build an astronomical telescope.

Equipment

Optical bench, several converging lenses, several diverging lenses, a ruler, an illuminated object with holder, a screen with holder, two lens holders.

Discussion and Theory

The *principal axis* of a lens is defined as the line which is perpendicular to the plane of the lens and goes through the center of the lens, as shown in Figures 33-1.

Rays of light will be *refracted* (bent) as they pass through a lens. A *converging lens* will always bend incoming rays parallel to the principal axis towards a point on the principal axis called the *focal point*, F, as shown in Figure 33-1. The distance from the focal point to the center of the lens is called the *focal length*, f. In a converging lens the focal length is always taken as being positive. A converging lens is always thicker at its center then along its rim.

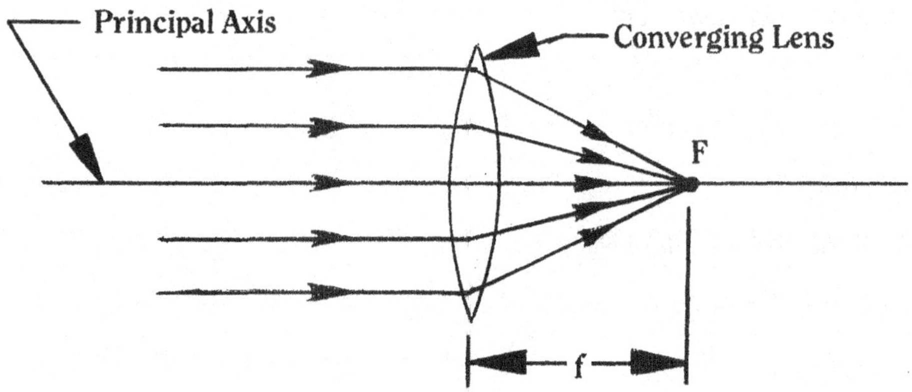

Figure 33-1 The Focal Point and Focal Length of a Converging Lens

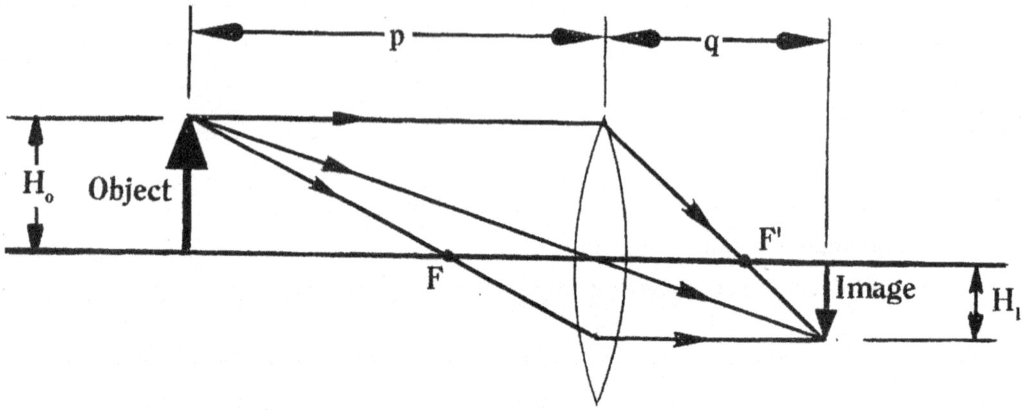

Figure 33-2 The Object and Image Distances of a Converging Lens

When an object is placed in front of a lens, at a point other then the focal point, an image is formed, as shown in Figure 33-2. The distance from the object to the center of the lens along the principal axis is called the *object distance,* p. The distance from the center of the lens to the image along the principal axis is called the *image distance,* q.

Figure 33-2 shows the path of several rays from the top of an object to the image for a converging lens, where the focal point falls between the object and the lens. The image is *inverted*. The rays actually converge and form a *real image*. A real image can be projected onto a screen.

The relationship between the object distance, p, the image distance, q, and the focal length, f, of the lens is given by the *lens equation*;

$$\frac{1}{p} + \frac{1}{q} = \frac{1}{f} \tag{33-1}$$

Solving for the image distance, q, yields,

$$q = \frac{f\,p}{p - f} \,. \tag{33-2}$$

The *magnification*, M, of a lens is defined as the image height, H_I, divided by the object height, H_0. Object and image heights are positive if erect; negative if inverted. The magnification can be shown also to equal the negative of the image distance, q, divided by the object distance, p. Therefore, we may write:

$$M = \frac{H_I}{H_0} = -\frac{q}{p}. \qquad (33\text{-}3)$$

Example

A 3.0 cm tall object is placed 25 cm to the left of a converging lens with a focal length of 10 cm. The image distance is obtained from equation (33-2) with p = 25 cm and f = 10 cm;

$$q = fp/(p - f) = 10(25)/(25\text{-}10) = 16.7 \text{ cm}.$$

The magnification is obtained from equation (33-3);

$$M = - q/p = - (16.7)/(25) = -0.67.$$

The image height is therefore,

$$H_I = H_0 M = (3.0)(-.67) = -2.0 \text{ cm}.$$

This image is 16.7 cm to the right of the lens. It is real and 2.0 cm tall. The negative image height tells us the image is inverted.

In optical instruments, several lenses are often used together. In an *astronomical telescope* two converging lenses are placed in series, as shown in Figure 33-3. The lens with the smaller focal length is called the *eyepiece*, and the lens with the larger focal length is called the *objective* lens. The distance between the two lenses is equal to the sum of the focal lengths of the eyepiece and the objective lens. Therefore, when the telescope is aimed at a distant object, the image from the objective lens is in front of the eyepiece at one of its focal points. The image of the objective lens is the object of the eyepiece. This causes parallel rays to emerge from the eyepiece which are then focused by the lens of the eye.

The magnification of a telescope is defined as:

$$M = \frac{\text{focal length of the objective lens}}{\text{focal length of the eyepiece}}$$

$$= \frac{f\,(\text{obj})}{f\,(\text{eye})} \qquad (33\text{-}4)$$

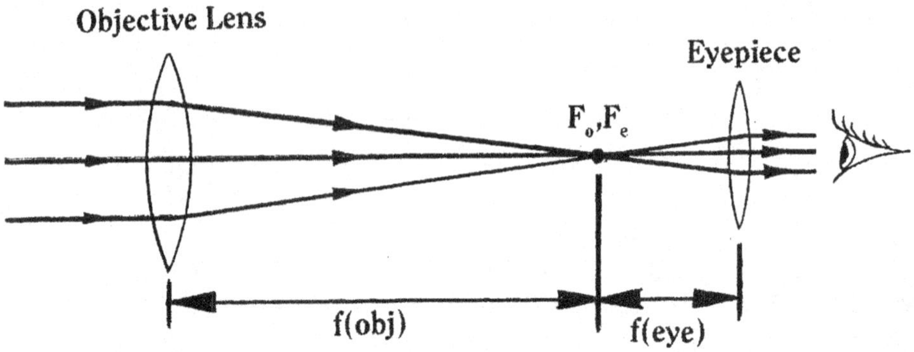

Figure 33-3 An Astronomical Telescope

Procedure

1. To find the focal point of a converging lens:

   (a) Point the optical bench towards a window so that the side with the zero end of the scale is towards the window.

   (b) Place a lens in a lens stand at the zero mark on the optical bench.

   (c) Place a screen in a stand on the bench in back of the lens away from the window.

   (d) Move the screen back and forth until the clearest image of a far away object is obtained.

   (e) Determine the distance between the lens and the screen. This is the measured focal length of the lens. Record this data in Table 33-1.

   (f) Repeat the procedure for all the lenses that are supplied. Record your results.

2. To apply the lens equation:

   (a) Place a lens with an approximate focal length of 10 cm at the 50 cm mark of the optical bench.

   (b) Place the object at the 35 cm mark. Determine the object distance? Record in Table 33-2.

(c) Place the screen beyond the 50 cm mark. Move this screen until the sharpest image is obtained. Determine the image distance, q, between the image and the lens and record in Table 33-2.

(d) Determine the theoretical image distance, q, from equation (33-2). Record in Table 33-2.

(e) Find the percent error in your experimental image distance. Record in Table 33-2.

(f) Measure the object height, H_0. Record in Table 33-2.

(g) Measure the image height, H_I. H_I is negative if it is inverted. Record in Table 33-2.

(h) Determine the magnification $M = H_I / H_0$. Record in Table 33-2.

(i) Determine the magnification $M = - q / p$. Record in Table 33-2.

(j) Place the object at the 30 cm mark.

(k) Using the same lens repeat steps (c) through (i). Record all your data in Table 33-2.

(l) Place the object at the 20 cm mark.

(m) Using the same lens repeat steps (c) through (i). Record all your data in Table 33-2.

(n) Your instructor may want you to repeat this experiment for other lenses. If so, you will be told which lenses and object distances to use.

3. To build an astronomical telescope.

(a) Point the optical bench towards the window.

(b) Place two lenses on the optical bench. From your lenses select the lens with the largest focal length and the lens with the smallest focal length. Of these two lenses, the lens with the larger focal length (the objective lens) should be between the window and the other lens (the eyepiece). Record the focal lengths of these lenses in Table 33-3.

(c) While moving the objective lens backwards and forward, look into the eyepiece through the objective lens until a clear image is formed. Is the image erect or inverted? Record your observations in Table 33-3.

(d) Determine the distance between the two lenses. Record.

(e) Add the focal lengths of the two lenses together. Does this length equal the distance between the lenses you just measured?

(f) Calculate the theoretical magnification of your telescope from equation (33-4) and record in Table 33-3.

(e) Make a second telescope using two different lenses. Record the focal length of the lenses used. Determine the actual distance between the lenses and the sum of the focal points. Calculate and record the magnification of this telescope in Table 33-3.

(f) From your observations which telescope gives a larger image? Indicate your observation in the space provided in Table 33-3. Is this consistent with what you would expect from your calculated magnifications?

Questions

1. Which of the following statements are true:

   (a) Lenses have two focal points.

   (b) A lens cannot have a flat side

2. Show how equation (33-2) is obtained by solving equation (33-1).

Name ______________________ Date Performed ______________________

Lab. Partners ______________________ Date Submitted ______________________

____

____

Table 33-1
Determination of the Focal Length of Several Lenses

	Given Value of f (cm)	Measured Value of f (cm)	Percent Error
Lens #1			
Lens #2			
Lens #3			
Lens #4			

Table 33-2
Applying the Lens Equation and Determining the Magnification

Focal Length (cm)	Object Distance (cm)	Image Distance (Experiment) (cm)	Image Distance (Theory) (cm)	% Error	Object Height H_O (cm)	Image Height H_I (cm)	M H_I/H_O	M $-q/p$

Table 33-3
Observing the Magnification of an Astronomical Telescope

	f(obj) (cm)	f(eye) (cm)	f(obj) + f(eye) (cm)	Distance Between Lenses (cm)	Erect or Inverted	Magnification f(obj)/f(eye)
Telescope #1						
Telescope #2						

Larger image observed for telescope # ________

EXPERIMENT 34

KEPLER'S LAWS

Objective

To study Kepler Three Laws

Equipment

A ruler with 0.1 centimeter divisions.

Discussion and Theory

Kepler's First Law

Kepler's First Law states that *the planets move in elliptical orbits around the sun with the sun at one focus.*

An ellipse has two points called *foci* (singular is *focus*) as shown in Figure 34-1. *The sum of the distances from any point on the ellipse to each focus is a constant.* Therefore, for the points P and Q on the ellipse in Figure 34-1

$$PF_1 + PF_2 = QF_1 + QF_2. \tag{34-1}$$

We will apply this property to the orbit of the hypothetical planet in orbit about the sun at focus F in Figure 34-5 to verify that the path is elliptical.

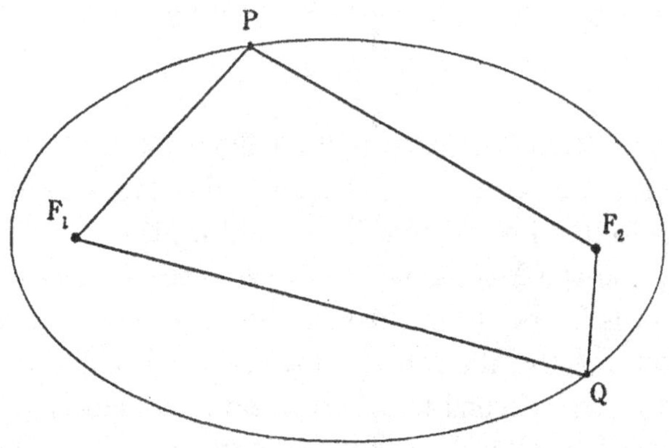

Figure 34-1 The Sum of the Distances from a Point on the Ellipse to each Focus is a Constant

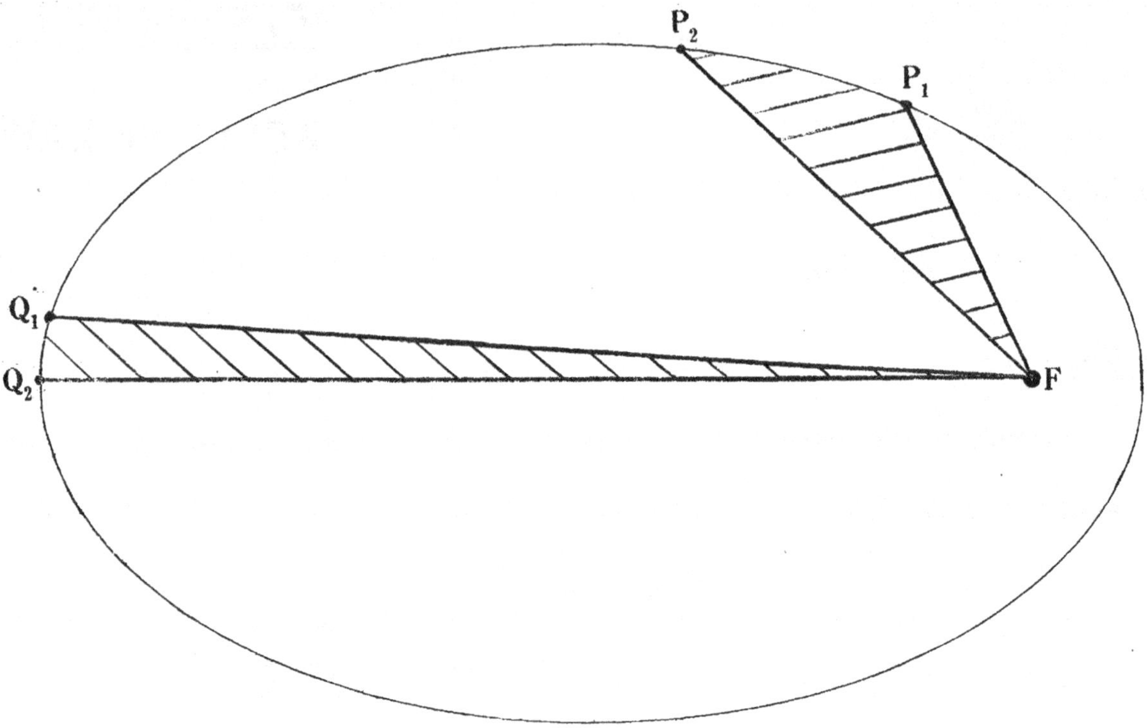

Figure 34-2 Kepler's Second Law

Kepler's Second Law

Kepler's Second Law states that *the straight line joining the sun and a planet sweeps out equal areas in equal intervals of time.*

Figure 34-2 shows the motion of a planet in orbit about the sun. As stated by Kepler's First Law, the orbit is elliptical and the sun is at focus F. If the planet takes the same amount of time to travel from point P_1 to point P_2 as it does to travel from point Q_1 to point Q_2, then according to Kepler's Second Law, the areas of segments P_1FP_2 and Q_1FQ_2 must be the same. The arc length P_1P_2 is longer then the arc length Q_1Q_2, where the planet is further from the sun. Since it takes the same amount of time for the planet to travel along both arc lengths, the planet must be traveling faster along the longer arc length, P_1P_2. Therefore, we see that Kepler's Second Law tells us how the velocity of a planet varies as it travels around the sun. A planet travels faster the closer it is to the sun and slower the further it is from the sun.

To verify Kepler's Second Law we will work with the motion of the hypothetical planet in orbit around the sun shown in Figure 34-6. The elliptical path of this planet has been divided into twenty segments by twenty points in such a way that the time for the planet to travel between neighboring points is the same. We will show that the area swept out by the line connecting the planet and the sun must be the same between neighboring points on the ellipse.

The semi-major axis, a, of an ellipse is half the length of its longest axis. The semi-minor axis, b, is half the length of its shortest axis. In the ellipse in Figure 34-6, the semi-major axis is measured from the center of the ellipse, point O, to point A. The semi-minor axis is measured from the center of the ellipse, point O, to point R. The area, A, of an ellipse is calculated from the formula

$$A = \pi ab. \tag{34-2}$$

Since the ellipse is divided into twenty segments, then the area, A', of each segment should be

$$A' = (1/20)\pi ab. \tag{34-3}$$

To calculate the areas of the individual segments formed by neighboring points on the ellipse we will approximate each segment with a triangle. The area of a triangle is found from

$$\text{Area} = (1/2)\ bh \tag{34-4}$$

where b is the base and h is the altitude. In Figure 34-3, segment GFE has been approximated by triangle GFE, by drawing a line between points G and E. The base, b, is the length GF, and the altitude, h, is the length of the line drawn from E perpendicular to the base.

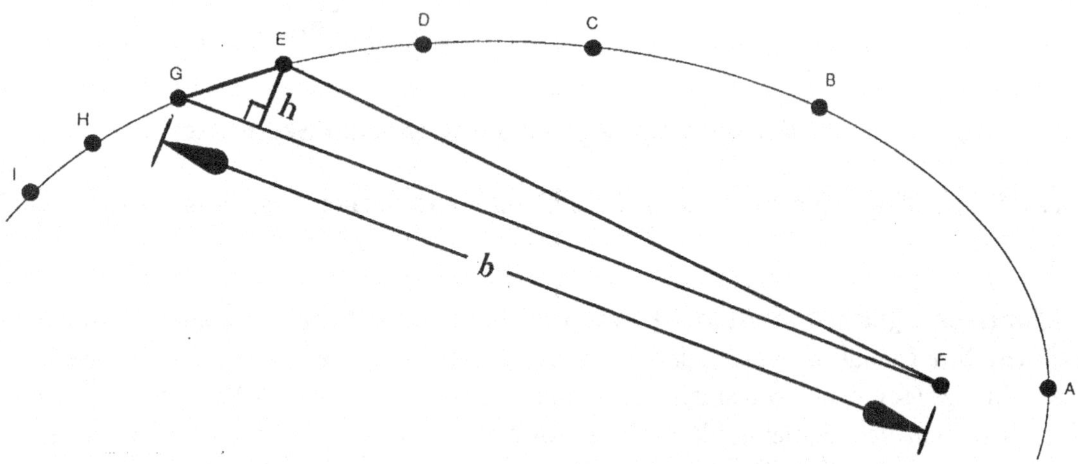

Figure 34-3 Measuring the Base and Altitude to Determine the Area of a Segment

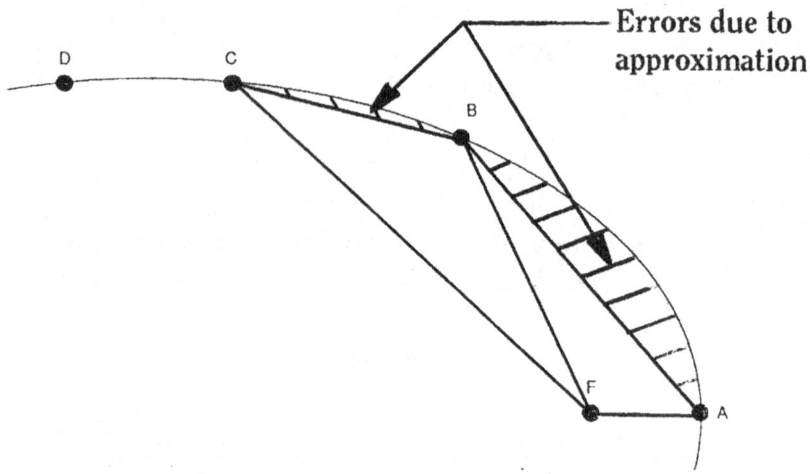

(a) Error in Areas Near Point F

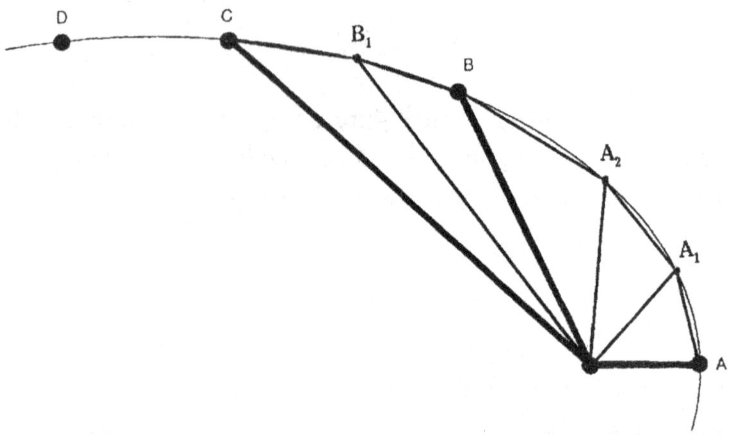

(b) Breaking Up Segments into Smaller Segments

Figure 34-4 Dividing Segments Near the Focal Point F into Smaller Segments to Calculate the Areas

For those segments closest to the focal point F, the error due to the approximation of the area of each segment by a triangle becomes unacceptable, as shown in Figure 34-4(a). To decrease this error, we will break up these segments into smaller segments as shown in Figure 34-4(b). We will then determine the areas of the smaller segments. We will approximate the area of segment CFB with the sum of the areas of triangles CFB_1 and B_1FB, and we will approximate the area of segment BFA with the sum of the areas of triangles BFA_2, A_2FA_1, and A_1FA.

Kepler's Third Law

Kepler's Third Law states that *for all the planets in the solar system, the quantity*

$$\frac{(\textit{Period of Revolution})^2}{(\textit{Average Distance from the Sun})^3} \tag{34-5}$$

is a constant.

An *astronomical unit* is the average distance of the earth from the sun. The average distance from the sun in astronomical units and the period of each planet in earth years will be used to calculate the quantity (34-5). According to the Kepler's Third Law the calculated values should be the same for all the planets.

Procedure

Kepler's First Law

To verify that the orbit represented by Figure 34-5 is elliptical do the following:

1. Measure the distances of point A from foci F and F' in Figure 34-5. Add these two distances together and record in Table 34-1.

2. Repeat step 1 for points B, G , H , K and any other point on the ellipse. Discuss your results in your conclusion.

Kepler's Second Law

Figure 34-6 is identical to Figure 34-5. To verify that the areas of all segments are equal do the following on Figure 34-6:

1. Draw a line from F to A, from F to B, from F to C, etc. on the ellipse in Figure 34-6. Stop with line FL.

2. From point K draw a line to point L to form triangle KFL. From point K drop a perpendicular to line FL. This perpendicular is the altitude of the triangle.

3. Measure the length of the base, FL, of triangle KFL. Record in Table 34-2.

4. Measure the length of the altitude of triangle KFL.. Record in Table 34-2.

5. Calculate the area of triangle KFL from equation (34-4). Record in Table 34-2.

6. Repeat steps 2, through 5 for segments JFK through CFD. For each segment enter the base, altitude and calculated area in Table 34-2.

7. Divide segment BFC into 2 segments, as shown in Figure 34-4(b), such that the area of each segment can be approximated by a triangle. Find the area of each triangle. Add the individual areas to find the total area of BFC. Enter the required values in Table 34-2.

8. Divide segment AFB into 3 segments, as shown in Figure 34-4(b), such that the area of each segment can be approximated by a triangle. Find the area of each triangle. Add the individual areas to find the total area of segment AFB. Enter the required values in Table 34-2.

9. Measure the distance from the center of the ellipse, O, to point A. This is the semi-major axis, a. Record in Table 34-2.

10. Measure the distance from the center of the ellipse, O, to point R. This is the semi- minor axis, b. Record in Table 34-2.

11. Calculate the area of the ellipse from equation (34-2). Record in Table 34-2.

12. Find $1/20^{th}$ of the area of the ellipse. Record in Table 34-2.

13. Use the value obtained in step 12 as the standard value and find the percent error in the calculated areas for each segment. Record these in Table 34-2.

Kepler's Third Law

To verify Kepler's Third Law do the following:

1. Table 34-3 lists the average distance form the sun and the period of revolution of all the planets in the solar system. For each planet calculate

$$\frac{(\text{Period of Revolution})^2}{(\text{Average Distance from the Sun})^3}$$

2. Discuss your results in your conclusion.

Questions

1. If figure 34-5 represents the motion of a planet about the sun, and if it takes 10 days for the planet to go from D to E, then how long does it take for the planet to go (a) from point A to point H? (b) once around the ellipse (i.e. from point A around the ellipse and back to point A)?

2. For the motion of the planet in Figure 34-5 at what point does the planet move the fastest, (b) the slowest?

3. A circle may be thought of as being an ellipse with the foci F and F' coincident. Since the radius of a circle is a constant, what could you conclude about the speed of a planet in a circular orbit about the sun from Kepler's Second Law?

4. From the Table 34-3, if a planet existed between Mars and Jupiter (a) the period of revolution would have to be more than how many years? (b) less than how many years?

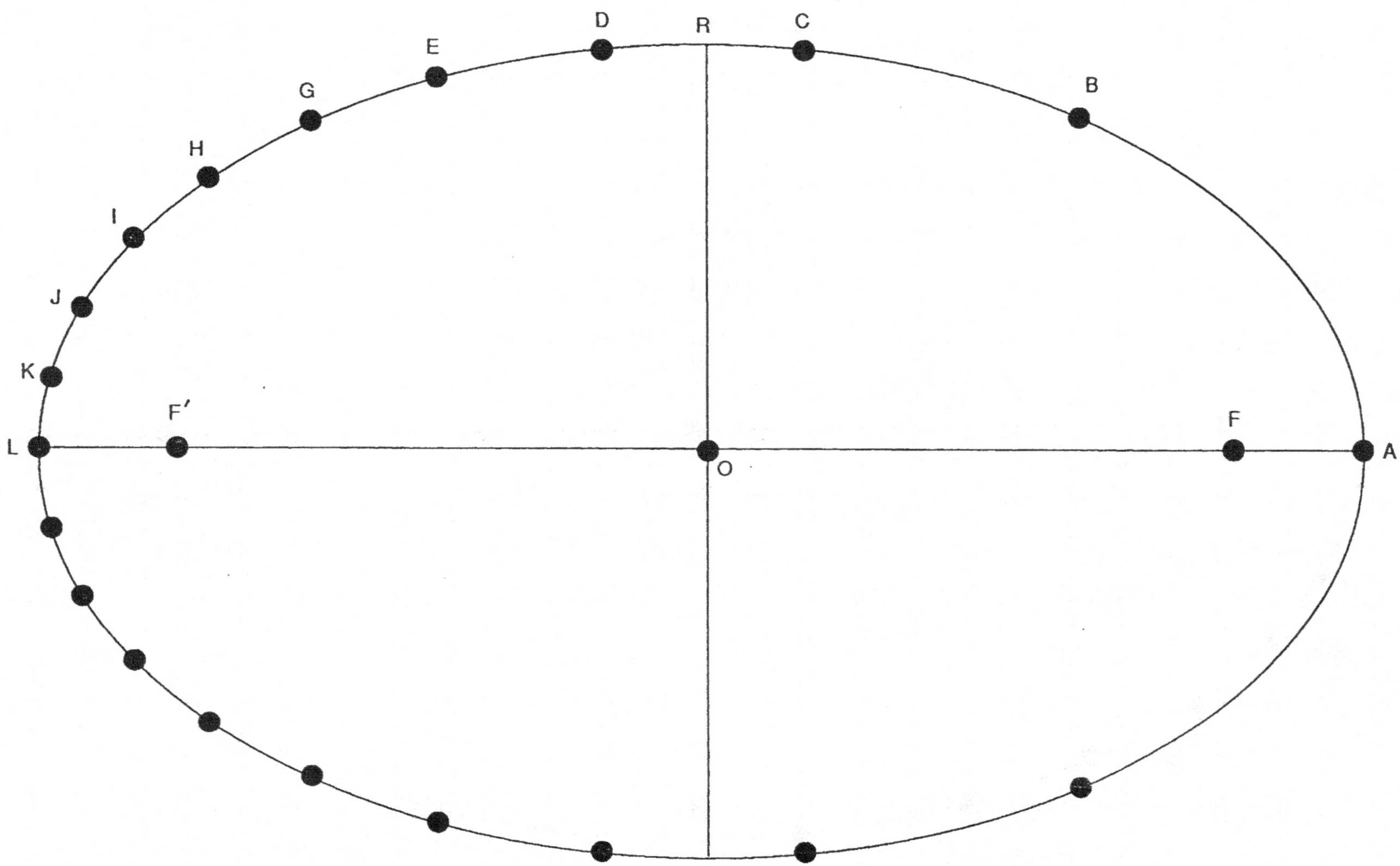

Figure 34-5 The Elliptical Orbit of a Hypothetical Planet

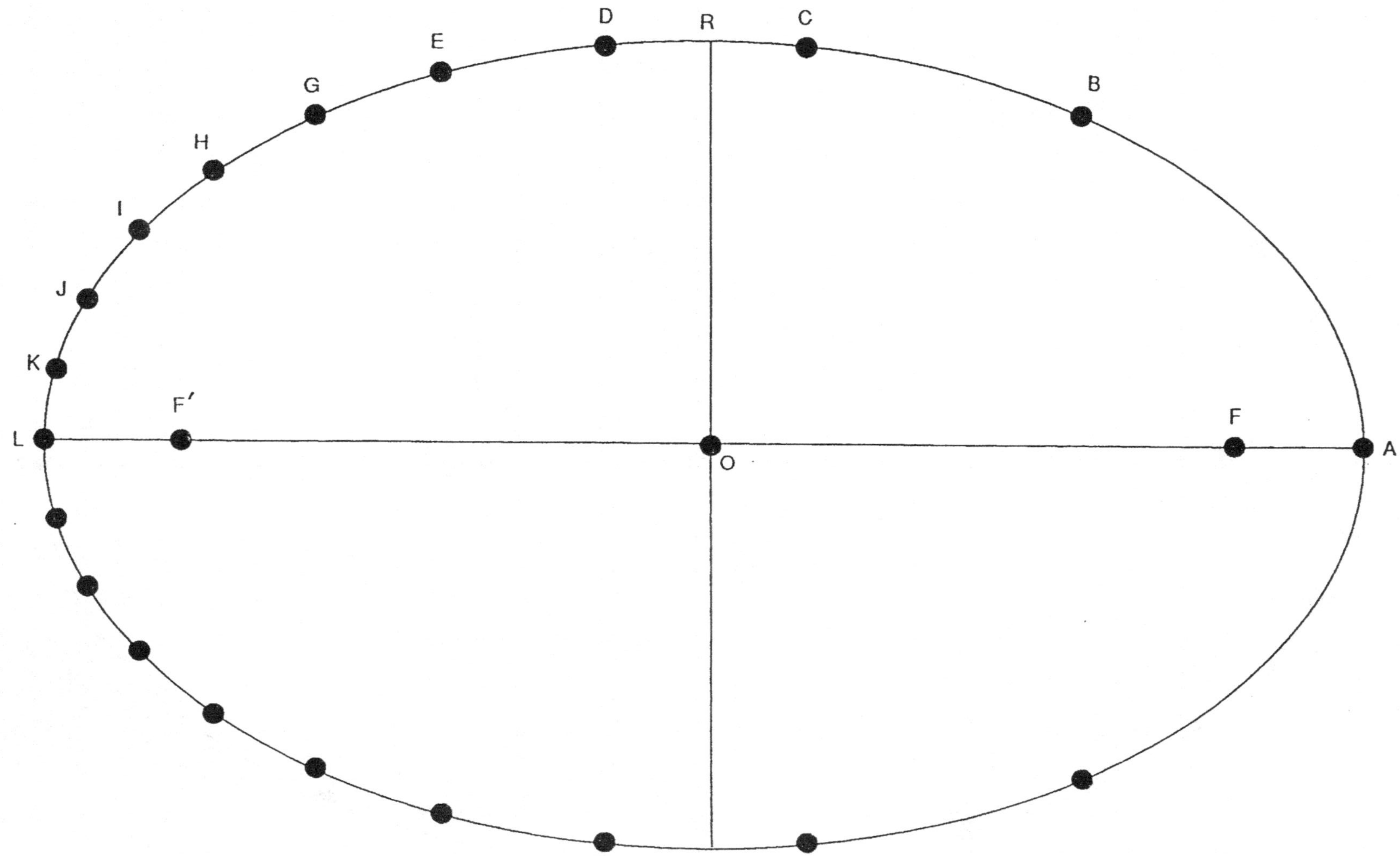

Figure 34-6 The Elliptical Orbit of a Hypothetical Planet

Name____________________________ Date Performed________________

Date Submitted________________

Table 34-1

Kepler's First Law: Verification that the Orbit in Figure 34-5 is Elliptical

	A	B	G	H	K	Any other point
Distance from F						
Distance from F'						
Sum						

Table 34-2

Kepler's Second Law: Verification that the Areas of the Segments are Equal

Triangle	Base (cm)	Altitude (cm)	Area (cm^2)	% Error
KFL				
JFK				
IFJ				
HFI				
GFH				
EFG				
DFE				
CFD				
BFC				
		Sum of two areas:		
AFB				
		Sum of three areas:		

a = ________ b = ________ πab = ________ $(1/20)\pi ab$ = ________

Name ______________________

Table 34-3
Kepler's Third Law: Verification that (Period)²/(Ave. Distance)³ = Constant

	Average Distance from Sun (Asrtonomical Units)	Period of Revolution (Earth Years)	$\frac{(\text{Period of Rev.})^2}{(\text{Ave. Distance from Sun})^3}$ (Year)²/(Astro. Units)³
Mercury	0.39	0.24	________
Venus	0.72	0.62	________
Earth	1.00	1.00	________
Mars	1.52	1.88	________
Jupiter	5.20	11.9	________
Saturn	9.54	29.5	________
Uranus	19.2	84.0	________
Neptune	30.1	165.	________
Pluto	39.4	248.	________

EXPERIMENT 35

LOCATING STARS AND OBSERVING RETROGRADE MOTION

Objectives

1. To learn how to express the coordinates of stars
2. To learn how to measure the angular distances between stars
3. To observe the retrograde motion of Mars

Equipment

Celestial globe, paper, soft crayon or washable felt tip pen.

Discussion and Theory

Before one can use the celestial globe, the following terms must be understood. Referring to a celestial globe will help in understanding these terms.

Celestial Sphere - An imaginary sphere encircling the earth upon which are all the stars. See Figure 35-1

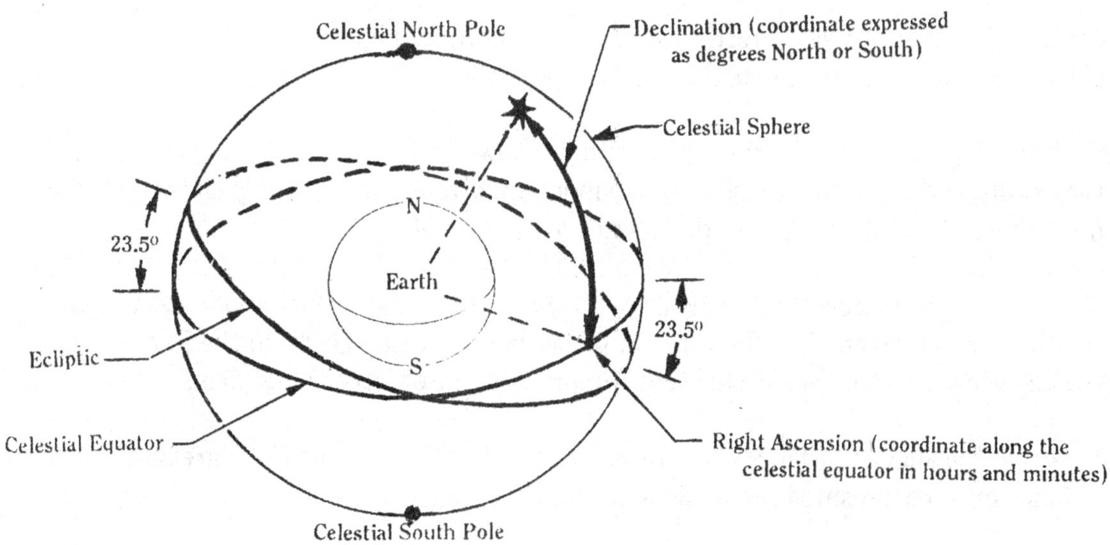

Figure 35-1 The Celestial Sphere

Celestial Poles - Imaginary points on the celestial sphere directly over the north and south poles of the earth as shown in Figure 35-1

Celestial Equator - The great circle on the celestial sphere 90° from the celestial poles. The celestial equator is directly over the equator of the earth as shown in Figure 35-1.

Right Ascension - The scale along the celestial equator which measures distance eastward in hours and minutes as shown in Figure 35-1. The range of right ascension is from 0 hr. to 24 hrs. Each hour of right ascension is divided into 60 minutes. Each minute is divided into 60 seconds. Right ascension on the celestial globe is similar to longitude on the terrestrial globe. In one hour of time, the right ascension of the celestial sphere increases by one hour.

Declination - The scale which measures distance in degrees north and south from the celestial equator as shown in Figure 35-1. A star above the celestial equator has north declination, a star below the celestial equator has south declination. North declination is also said to be positive (+); south declination is said to be negative (-). On the laboratory globes, the declinations are indicated every 15° with 5° subdivisions on the great circle that passes through the celestial poles and goes through 0 and 12 hours of right ascension. Declination on the celestial globe is similar to latitude on a terrestrial globe.

Ecliptic - The circle defined by the apparent path of the sun in the sky on the celestial globe. The ecliptic is inclined 23 1/2° to the celestial equator as shown in Figure 35-1. The dates written on the ecliptic of a celestial globe give the location of the sun on the celestial sphere during the year.

Zenith - The point in sky directly overhead as shown in Figure 35-2. On the celestial globe it is the point at the top of the celestial sphere.

Horizon - A great circle 90° from the zenith as shown in Figure 35-2. Only that part of the sky above the horizon is visible at any moment in time. At night only stars above the horizon may be seen; stars below the horizon are not visible.

Meridian - A great circle that passes through the north and south celestial poles and the zenith of the observer. On a the celestial globe the metal ring at 90° to the horizon and circling vertically over the celestial sphere represents the observer's meridian.

Altitude - The angular distance above or below the horizon as shown in Figure 35-2. This distance must be measured perpendicular to the horizon.

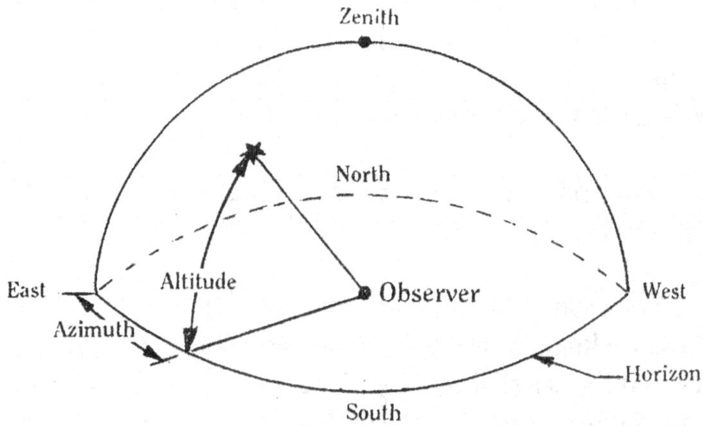

Figure 35-2 The Altitude and Azimuth of a Star

Azimuth - The angle in degrees north or south along the horizon measured relative to the east and west as shown in Figure 35-2 (e.g., 15° north of east, or 21° south of west).

Constellation - A configuration of stars named for a particular object, person or animal. It is identified on the celestial globe with all capital letters. All the stars in a particular constellation are joined together with lines.

Retrograde Motion - An apparent reversal in the direction of motion of a planet on the celestial sphere that occurs when the earth passes by it as it goes around the sun.

Procedure

The Location of Stars in the Sky

To locate stars in the sky, two coordinates are needed, the declination and right ascension. Declination is measured in degrees. Right ascension is measured in hours and minutes.

1. Make a ruler to measure angular distances on your celestial sphere as follows:

a) Take a piece of paper, fold it several times so that you get a strip about 1/2 to 1 inch wide and about 10 inches long.

b) Look along the celestial equator for the right ascension 0 hrs (or 12 hrs). Notice the great circle which is at right ascension 0 hrs (or 12 hrs) and goes through the north and south poles. On this great circle, there is a scale of declination every 5° from 0 to 75°.

c) Place your paper against this scale and carefully copy the scale onto your paper. This will be your ruler which you will use to measure angular distances.

2. Measure the declination and right ascension of each star listed in Table 35-2 and indicate the constellation in which it is located as follows:

a) The declination of a star is the angular distance north or south of the celestial equator. To use your ruler to measure the declination, place the scale perpendicular to the celestial equator. The zero mark of the scale must be on the celestial equator. Measure the number of degrees north or south to the star. The number of degrees north or south is the declination.

b) To find the right ascension, read the hours and estimate the minutes along the celestial equator.

c) To find the constellation, look for the capitalized word near all the stars that are connected with lines that contain the star you are looking at.

Example

To find the coordinates for the star Altair. Use your scale as explained above to find the declination. You should find it to be approximately 8° N. To find the right ascension, notice that it is between 19 and 20 hrs. Your estimate of the minutes should be about 50 minutes (about $5/6^{th}$ of the way from 19 to 20 hrs) Finally, notice that the constellation it is in is AQUILA (all the letters of a constellation on the globe are capitalized).

To Measure Distances Between Stars

Distances between stars are measured in angles. Put the zero mark of your ruler in the center of one star and read the angular distance to the center of the other star. Go to Table 35-3 and fill in the data according to the instructions.

To List the Constellations of the Zodiac

Start at right ascension 0 hr, 0 min and follow the ecliptic around 24 hours until you reach 0 hr, 0 min again. Name and list all the constellations of the zodiac in their proper order in Table 35-4

To Observe the Motion of the Planet Mars Along the Celestial Sphere

There are no planets printed on the celestial globe. There is, of course, a reason for this. The planets cannot be printed on the celestial globe because, unlike the stars which are fixed, the planets are constantly changing their positions in the heavens. They are wanderers, as the name planet implies. In Table 35-1, the coordinates for Mars are tabulated for a two-year period. Since the planet moves slowly, the data is given at intervals of one month or less. Plot each of the coordinates on the celestial globe. Use either a soft crayon or a washable felt tip pen that leaves a washable stain (Do NOT use a sharp pen that can mar the surface). As each point is plotted connect it to the previously plotted point with a line.

After you have plotted your data from Table 35-1, describe in a well written paragraph your observations concerning the motion of Mars.

Table 35-1
The Retrograde Motion of Mars

Elapsed Time Days	Right Ascension hrs	min	Declination degrees	Elapsed Time Days	Right Ascension hrs	min	Declination degrees
1	0	9	0	349	11	17	9N
30	1	34	9N	365	10	55	11N
61	3	3	17N	380	10	46	11N
91	4	32	22N	395	10	47	10N
122	6	4	24N	410	10	57	8N
153	7	32	23N	426	11	16	6N
183	8	49	19N	441	11	37	3N
214	10	0	14N	456	12	4	0
229	10	28	12N	487	13	6	7S
244	10	58	9N	518	14	19	15S
259	11	20	7N	548	15	41	21S
275	11	42	5N	579	17	16	24S
290	11	54	4N	609	18	56	24S
306	11	58	4N	640	20	38	20S
321	11	52	5N	671	22	15	12S
334	11	37	9N	699	23	37	3S

Name______________________________ Date Performed____________________
Lab Partners ______________________ Date Submitted____________________
____
____

Table 35-2
Locating Stars

Star	Declination	Right Ascension	Constellation
1. Altair			
2. Deneb			
3. Vega			
4. Capella			
5. Sirius			
6. Betelgeuse			
7. Rigel			
8. Aldebaran			
9. Arcturus			
10. Procyon			
11. Spica			
12. Castor			
13. Pollux			
14. Antares			
15. Canopus			

Name________________________________

Table 35-3
Measuring Distances Between Stars

The Northern Triangle consists of the stars Deneb, Altair and Vega. Measure the distances between these stars.

Deneb - Altair ______________________________
Deneb - Vega ______________________________
Vega - Altair ______________________________

Measure the distances between the following stars:

Capella - Rigel ____________________________
Castor - Pollux ____________________________
Procyon - Pollux ____________________________
Procyon - Castor ____________________________

Find the Constellation Pegasus. In this Constellation there is a "great square". Calling the stars A,B,C, and D as shown in the sketch measure the distances:

A - B ______________
B - C ______________
C - D ______________
D - A ______________

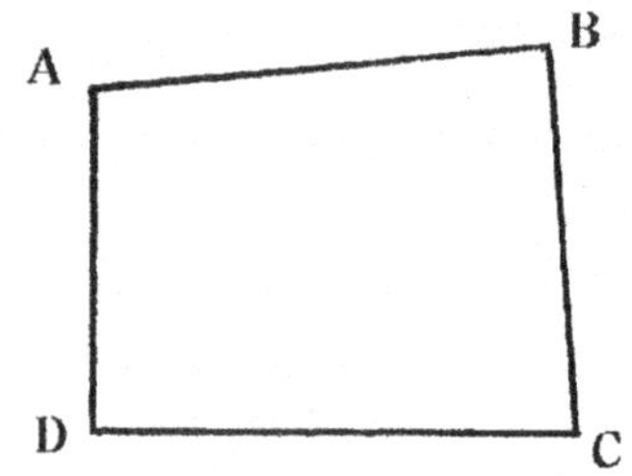

Table 35-4
Constellations of the Zodiac

1. ________________________ 7. ________________________
2. ________________________ 8. ________________________
3. ________________________ 9. ________________________
4. ________________________ 10. ________________________
5. ________________________ 11. ________________________
6. ________________________ 12. ________________________

Describe the retrograde motion of Mars. ____
____
____
____
____

EXPERIMENT 36

THE LENGTH OF A DAY

Objective

1. To study the variation in the length of the day with the seasons at 41° N latitude

2. To study how the length of the day varies with latitude on March 21, June 21, September 21, and December 21.

Equipment

Celestial globe, paper, soft crayon or soft washable felt tip pen.

Discussion and Theory

The objects in the sky that are visible depend upon several factors: (a) the latitude of the observer's location (e.g. 41° North latitude at New York City), (b) the month and day of the year, and (c) the time of day or night.

Before one can use the celestial globe to study variations in the length of the day, it is necessary to know how to set the celestial globe so that it is properly oriented for a particular latitude. A globe set for a particular latitude is said to be *rectified*.

Rectifying the Globe for Latitude

Notice that the stand of the globe indicates the directions North, East, South, and West. Place the globe so that the North arrow is on the side of the globe away from you and the South arrow is facing you. To rectify the globe for North latitudes, take the North pole of the globe and push it down towards the North horizon until the scale on the meridian shows the value of the desired latitude at the very top of the meridian. An alternate way is to push the North pole down toward the North horizon until the reading on the North horizon is 90° - (the value of the desired latitude).

Examples

(1) To rectify the globe for 0° latitude (i.e., over the equator). Take the North Pole of the globe and push it all the way down to the horizon on the North side. The reading on the North side of the meridian will be $90^{\circ} - 0^{\circ} = 90^{\circ}$ and the reading at the very top of the meridian will be 0°. You can verify that the globe is rectified over the equator by looking down at the small terrestrial globe within the celestial globe. You will notice that the equator is under the Zenith.

(2) To set the celestial globe for a latitude of 10^0 N, push the North pole 10^0 up from the horizon. The reading of the meridian at the North horizon for a 10^0 North latitude setting should be $90^0 - 10^0 = 80^0$. Because the horizontal ring for the horizon has a 5^0 width, the readings will be 82.5^0 on the top of the ring for the horizon and 77.2^0 on the bottom, giving an average of 80^0 in the center of the horizon. The topmost point (Zenith) on the observer's meridian facing the South will be at 10^0 N.

(3) New York City is at 41^0 N latitude. To set the globe for this latitude, we must have a reading of $90^0 - 41^0 = 49^0$ on the meridian on the North horizon of the globe (The reading on the meridian just above the horizon will be 51.5^0 and the reading below will be 46.5^0). A reading of 41^0 N should be on the top of the meridian. If one looks down at the small terrestrial globe within the celestial globe, one should be able to turn the terrestrial globe so that New York City is at the top of the terrestrial globe.

Determination of the Length of a Day

To find the length of the day, on a particular date and particular latitude, it is necessary to find the time from sunrise to noon and the time from noon to sunset. The length of the day is the sum of these two times. To determine the length of a day on the celestial globe:

(a) Rectify the globe at the desired latitude.

(b) Find the desired date on the ecliptic. Mark its location with a washable felt tip marker. This gives location of the sun on the celestial sphere on the date indicated.

(c) Place the marked date (i.e. the sun) at the east horizon.

(d) Recall that it takes one hour for the celestial globe to turn one hour right ascension. Turn the celestial globe towards the vertical metal ring of the meridian. As the globe turns count the number of hours and minutes on the right ascension scale that pass under the meridian. The number of hours and minutes counted is the time it takes the sun to go from sunrise to noon.

(e) The determination of the time from noon to sunset is similar, except that you start with the sun under the meridian. You turn the globe until the sun reaches the west horizon and count the number of hours and minutes of right ascension that pass under the metal ring of the meridian. The number of hours counted is the time from noon to sunset.

(f) Add the time from sunrise to noon and noon to sunset to obtain the length of the day.

Example

Choose the day, May 1, at New York City (41^0 North Latitude). Rectify the globe for a latitude of 41^0 N. On the Ecliptic, put a small mark with a washable marking pen on May 1. This represents the sun (the small yellow globe can also be set to represent the sun on May 1). Place the celestial globe so that the South arrow on the stand faces you. Rotate the globe until the mark on May 1 is directly under the east horizon (i.e., it should be under the center of the horizontal ring on the side marked East). The reading of the right ascension scale under the meridian should be about 19 hours, 30 minutes. Now turn the celestial globe and count the number of hours and minutes that pass under the metal ring for the meridian, until the sun is under the meridian (the noon position of the sun). The reading of the right ascension scale under the meridian should be 2 hours, 30 minutes. The number of hours and minutes counted that passed under the meridian should be about 7 hours; this is the time it took for the right ascension to change from 19 hours, 30 minutes (sunrise) through 24 hrs (or 0 hrs) to 2 hours, 30 minutes (noon). Sunrise occurred at 5:00 A.M. (12:00 noon minus 7 hours).

To find the time to sunset from noon, start with the sun under the meridian (the right ascension reads 2 hours, 30 minutes under the meridian). Turn the celestial globe and count the number of hours and minutes that pass under the meridian, until the sun reaches the west horizon (middle of the horizontal metal ring on the side marked West). The right ascension should read about 9 hours, 30 minutes under the meridian. The number of hours and minutes counted that passed under the meridian should be about 7 hours; this is the time it took for the right ascension to change from 2 hours, 30 minutes (noon) to 9 hours, 30 minutes (sunset). Sunset occurred at 7:00 P.M. (12:00 noon plus 7 hours).

The length of the day is 14 hours (7 hours from sunrise to noon and 7 hours from noon to sunset.

Procedure

1. Rectify the globe at 41° North Latitude. Determine the l.a.t. (local apparent time-sundial time) of sunrise, the l.a.t. of sunset and the length of the day for New York City for the days shown in Table 36-2.

2. Determine the l.a.t., of sunrise and sunset and the length of day on June 21 for the latitudes in Table 36-2. Indicate where there is no sunrise and sunset by writing across the table at the appropriate latitude "sun is always above the horizon - 24 hours of daylight", or "sun is always below the horizon - 24 hours of night".

3. Determine the l.a.t. of sunrise and sunset and the length of day on December 21 for the latitudes in Table 36-3. Indicate where there is no sunrise and sunset by writing across the table for the appropriate latitude "sun is always above the horizon - 24 hours of daylight", or "sun is always below the horizon - 24 hours of night".

4. Determine the l.a.t. of sunrise and sunset and the length of day on March 21 for the latitudes in Table 36- 4.

5. Determine the l.a.t. of sunrise and sunset and the length of day on September 21 for the latitudes in Table 36- 5.

Questions

1. Add together the length of the day of June 21st with the length of the day on Dec. 21st. How many hours is this? What should this be?

2. Add the length of the day on Sept. 21st with the length on March 21st. How many hours is this? What should this be?

3. What is the length of the day and the length of the night on Sept. 21st and March 21st. Why are Sept. 21st and March 21st call equinoxes?

4. Above what latitude do you have 24 hours of night on Dec. 21st?

5. Above what latitudes do you have 24 hours of day on June 21st?

6. Our clocks are set to *standard time* in the winter and *daylight savings time* in the summer. Which of these is closer to local apparent time. Explain?

7. Compare the results for the variation in the length of a day with latitude for March 21 and September 21. Are there any similarities?

Name ______________________ Date Performed ______________________

Lab. Partners ______________________ Date Submitted ______________________

____

____

Table 36-1

Variation of the Length of the Day with Date at 41° N Latitude

Date	Time of Sunrise	Time of Sunset	Length of the Day
June 21			
Sept. 21			
Dec. 21			
March 21			
Today' Date			
Your Birthday			

Table 36-2

Variation in the Length of the Day with Latitude on June 21

Latitude	Time of Sunrise	Time of Sunset	Length of the Day
0° North			
23° North			
41° North			
67° North			
90° North			

Table 36-2

Variation in the Length of the Day with Latitude on December 21

Latitude	Time of Sunrise	Time of Sunset	Length of the Day
0° North			
23° North			
41° North			
67° North			
90° North			

Name ______________________

Table 36-4
Variation in the Length of the Day with Latitude on March 21

Latitude	Time of Sunrise	Time of Sunset	Length of the Day
0° North	________	________	________
23° North	________	________	________
41° North	________	________	________
67° North	________	________	________
90° North	________	________	________

Table 36-5
Variation in the Length of the Day with Latitude on September 21

Latitude	Time of Sunrise	Time of Sunset	Length of the Day
0° North	________	________	________
23° North	________	________	________
41° North	________	________	________
67° North	________	________	________
90° North	________	________	________

EXPERIMENT 37

THE MOTION OF THE SUN IN THE SKY

Objective

1. To learn how the altitude of the sun varies during the year at various latitudes and
2. To learn how the azimuth varies with the time of the year and latitude.

Equipment

Celestial Globe, paper, soft crayon or washable felt tip pen.

Discussion and Theory

To understand this experiment the student should review the following definitions: (a) The *altitude* of an object in the sky is the angular distance measured up (perpendicular) to the object from the horizon as was shown in Figure 35-2. (b) The *azimuth* is the angle along the horizon, as was shown in Figure 35-2. We will express the azimuth as the number degrees North or South of East, or the number degrees North or South of West.

The sun appears to travel through the sky along a circular path about an axis that runs from an observer through the north or south celestial pole. Figure 37-1 shows the path of the sun as it would appear at a middle latitude in the northern hemisphere. As shown, the maximum noon altitude of the sun varies by 47° (= 2 x 23.5°) over the course of the year. The smallest noon altitude occurs on the *winter solstice* (~Dec. 21). The noon altitude is 23.5° higher during an *equinox* (~March 21 or ~Sept. 21). The maximum noon altitude occurs on the *summer solstice* (~June 21) and is an addition 23.5° higher than during an equinox.

The azimuth of the sun during sunrise and sunset varies throughout the year. Figure 37-1 shows the extent of the variation of the azimuth during sunrise at a middle latitude. This azimuth is most southerly on the winter solstice and most northerly on the summer solstice.

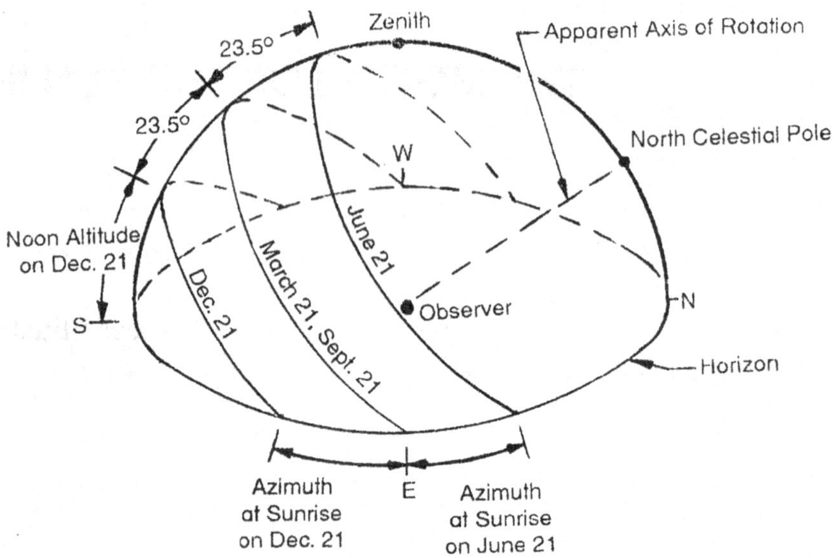

Figure 37-1 The Apparent Motion of the Sun at Middle Latitudes in the Northern Hemisphere

Determination of The Altitude

On the bottom of page 251, of the Procedure of Experiment 35, *The Location of Stars in the Sky,* step 1, a measuring device (flexible ruler) was made to measure the angular distances between the stars and to determine the declination of the stars. This same flexible ruler will be used to measure the altitude of the sun. To do this, place the flexible ruler on the globe so that it passes through the sun and through the zenith (topmost point on the globe) and is perpendicular to the horizon. The angular distance from the horizon to the sun is the altitude.

Determination of the Azimuth

In this experiment you will determine the azimuth of the sun during sunrise and sunset at various dates and for various latitudes. To make this measurement, place the flexible ruler on the globe at the horizon with the zero over due east or due west. Measure the azimuth along the horizon as the number of degrees towards the north or the south (i.e., 25° north of east).

Procedure

1. Rectify the globe for the latitudes indicated and mark the position of the sun with a washable felt tip marker for the indicated dates. Instructions on how to rectify the globe were give in the discussion of Experiment 36 on page 257. (To avoid rectifying the globe for the same latitudes over again you may wish to read ahead to steps 2,3 and 4 and collect all the applicable data while the globe is rectified at each latitude)

a. Rectify the globe for 41° North latitude. Determine the altitude of the sun in the sky as a function of time for each of the dates indicated. Plot the results for each date on graph #1. Label each curve with the appropriate date.
   1. June 21
   2. Dec. 21.
   3. March 21.

b. Rectify the globe for 0° North latitude. Determine the altitude of the sun in the sky as a function of time for each of the dates indicated. Plot the results for each date on graph #2. Label each curve with the appropriate date.
   1. June 21
   2. Dec. 21.
   3. March 21.

c. Rectify the globe for 90° North latitude. Determine the altitude of the sun in the sky as a function of time for each of the dates indicated. Plot the results for each date on graph #3. Label each curve with the appropriate date. If the sun is never above the horizon for any date, write on the top line of the graph, "There are 24 hrs of darkness on________"
   1. June 21
   2. Dec. 21.
   3. March 21.

d. Rectify the globe for 75° North latitude. Determine the altitude of the sun in the sky as a function of time for each of the dates indicated. Plot the results for each date on graph #4. Label each curve with the appropriate date. If the sun is never above the horizon for any date, write on the top line of the graph, "There are 24 hrs of darkness on________"
   1. June 21
   2. Dec. 21.
   3. March 21.

2. In step 1 above, you measured the altitude of the sun on June 21 at various latitudes. In Table 37-1 record the altitude of the sun at noon for each of the latitudes indicated.

2. Rectify your globe at 41° North. Use your flexible ruler to determine the azimuth at sunrise and sunset on June 21, Sept. 21, Dec. 21, March 21 and the day of the experiment. Record in Table 37-2.

3. Rectify your globe to the latitudes indicated in Table 37-3. Determine the azimuth for sunrise and sunset on June 21. Record in Table 3

4. If you are asked to answer question 1 through 3 do so before you leave the laboratory. You need the celestial globe.

Questions

1. At a latitude of 75° N, during what period of time does the sun never set.

2. Is there any latitude at which for one moment during the day of June 21 the Ecliptic lies along the Horizon? What time of day is it?

3 On the day of the experiment at 41° North latitude:

a. What Constellation in the Zodiac is rising at 9 P.M?
b. What Constellation in the Zodiac is setting at 9 P.M?
c. How many hours is Vega above the horizon?
d. How many hours is the Big Dipper above the horizon?
e. Which region in the sky is never above the horizon?

4. If you live at the North Pole, how much of the sky do you ever get to see?

5. Where on earth do you see the entire sky over the course of a year?

6. Does the azimuth of the sun at sunrise and sunset vary with latitude?

7. Does the azimuth of the sun at sunrise and sunset vary with the seasons?

8. Does the noon altitude of the sun vary with latitude?

9. The axis of the earth is tilted 23.5°. Between which latitudes does the sun reach an altitude of 90° sometime during the year?

Name ______________________________ Date Performed ______________________________
Lab Partners ______________________________ Date Submitted ______________________________
____
____

Graph 1 (41° N Latitude)

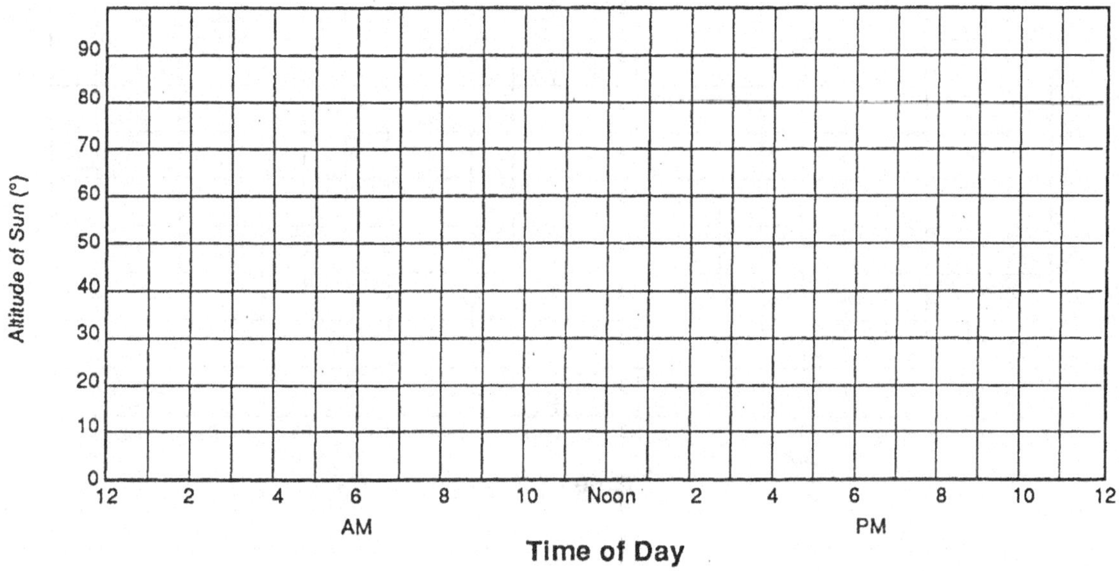

Graph 2 (0° N Latitude)

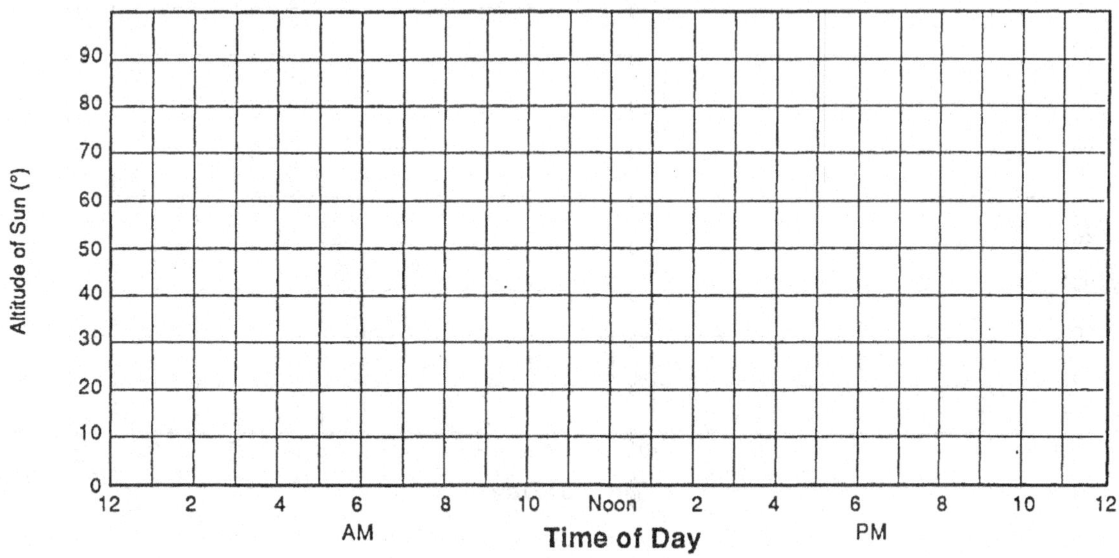

Name

Graph 3 (90° N Latitude)

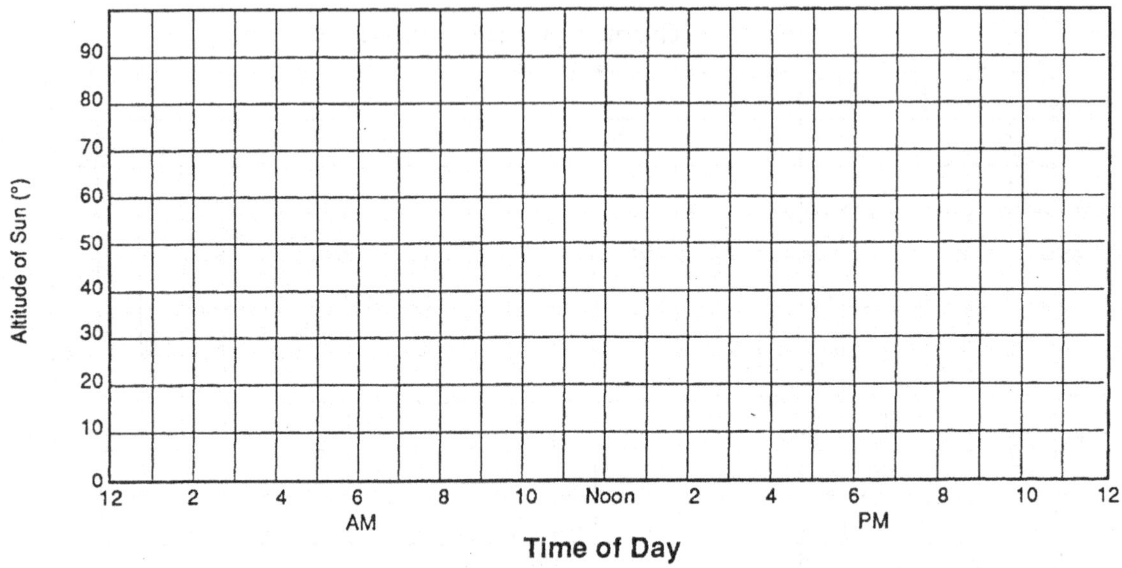

Graph 4 (75° N Latitude)

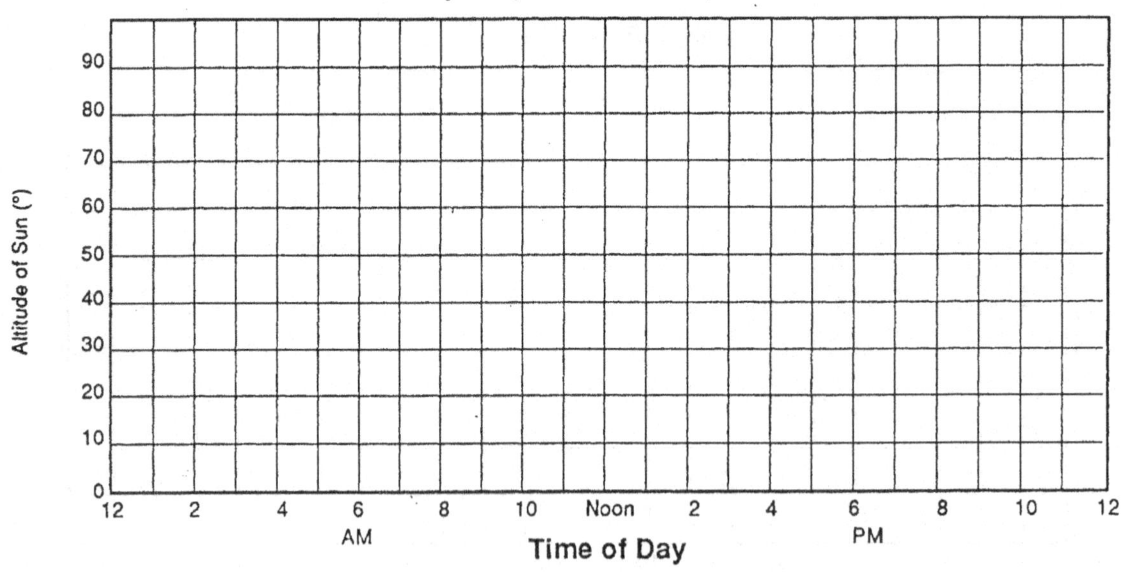

Name ____________________

Table 37-1
Altitude of the Noon Sun on June 21 at Various Latitudes

Latitude	Altitude
0°	________
41°	________
75°	________
90°	________

Table 37-2
Azimuth at 41° North at Various Dates

Date	Azimuth at Sunrise	Azimuth at Sunset
June 21	________	________
Sept. 21	________	________
Dec. 21	________	________
March 21	________	________
Today	________	________

Table 37-3
Azimuth at Sunrise and Sunset at Various Latitudes for June 21

Latitude	Azimuth at Sunrise	Azimuth at Sunset
0°	________	________
23°	________	________
41°	________	________
66°	________	________
90°	________	________

EXPERIMENT 38

OHM'S LAW

Objective

To study Ohm's Law

Equipment

Power supply, milliammeter, voltmeter, decade, resistance box.

Discussion and Theory

A basic relationship in electricity is that in a direct current (D.C.) circuit:

$$R = \frac{V}{I} \qquad (38\text{-}1)$$

where R is the resistance in Ohms, V is the potential difference in Volts and I is the current in Amperes. When the resistance is a constant, we have *Ohm's Law* where equation (38-1) can be rewritten as:

$$V = I\,R \qquad (38\text{-}2)$$

This equation is the form y = mx (a straight line with a zero y-intercept).

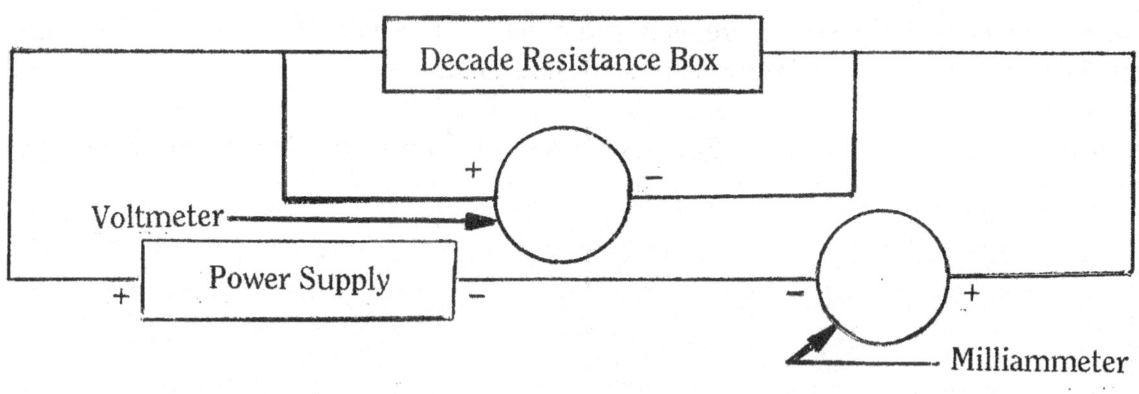

Figure 38-1 The Experimental Set-Up for Ohm's Law

Procedure

1. Assemble the parts as shown in the diagram in Figure 38-1. Do not plug the power supply into the electrical outlet until your instructor checks the circuit. Set the voltmeter on the 10 volt scale.

2. Set the decade resistance box to the value specified by your instructor.

3. *Slowly* adjust the power supply until the voltmeter reads one volt.

4. Read the milliammeter and record the value in Table 38-1.

5. Repeat steps 3 and 4 for voltmeter readings of 2 volts, 3 volts, 4 volts, 5 volts up to 8 volts.

6. Convert the milliammeter readings into Amperes and record in Table 38-1. One milliampere equals .001 Amperes.

7. Plot the data from Table 38-1. Plot the voltage, V, along the y axis and the current, I, in Amperes along the x axis.

8. Find the slope. This is your experimental value of the resistance, R. Record this value in Table 38-1.

9. Find the percent error between your experimental value and the standard value put on the decade resistance box in step 2.

Questions

1. A 10 Ampere current passes through a 5 ohm resistor. What is the potential difference (voltage) across the resistor?

2. When a potential difference of 100 volts is applied across a resistor a current of 2.0 Amperes is measured passing through the resistor. What is the resistance?

3. How much current will pass through a 10 ohm resistor if a potential difference of 20 volts is applied across it?

Name ______________________ Date Performed ______________________

Lab Partners ______________________ Date Submitted ______________________

____

____

Table 38-1
Ohm's Law

Voltage (Volts)	Current(Milliamperes)	Current (Amperes)
1.0	________	________
2.0	________	________
3.0	________	________
4.0	________	________
5.0	________	________
6.0	________	________
7.0	________	________
8.0	________	________

Slope = ________

Percent Error = ________

EXPERIMENT 39

SERIES CIRCUITS

Objective

To study the relationship between the currents and voltages in a series circuit.

Equipment

Power supply, a milliammeter, a voltmeter, an ohmmeter, a resistance board, leads, a decade resistance box.

Discussion and Theory

In this experiment we will study the resistors in the series circuit shown in Figure 39-1. The symbol 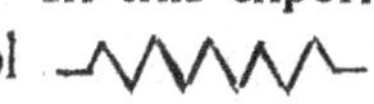 is used to indicate a resistance. We will verify that:

1. The potential difference across the power supply equals the sum of the drops in the potential difference across each resistor. This is *Kirchhoff's 2nd Law* which states that *the sum of the changes in potential difference around any loop is equal to zero.*

2. The current in a series circuit is everywhere the same.

3. Several resistances in series may be replaced by one equivalent resistance, R_{eq}, equal to the sum of the resistances. Therefore, for three resistors, R_{AB}, R_{CD}, and R_{EF}, the equivalent resistance is

$$R_{eq} = R_{AB} + R_{CD} + R_{EF}. \qquad (39\text{-}1)$$

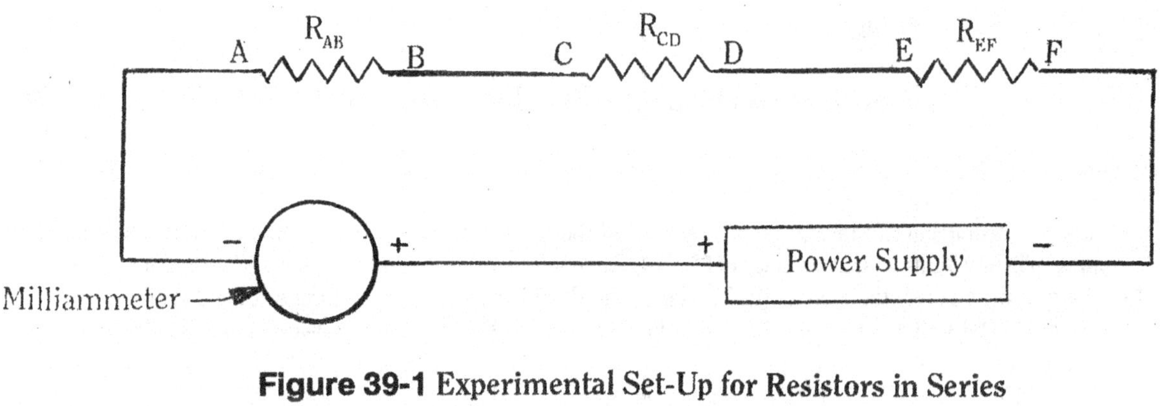

Figure 39-1 Experimental Set-Up for Resistors in Series

Procedure

1. Before you assemble the circuit, look at the resistance board which consists of three carbon resistors mounted on a wooden board. The ends of the resistors are connected to terminals labeled A, B, C etc.. The resistors are identified by the terminals at each end (e.g. R_{AB} is between terminals A and B). There are no numbers written on the resistances to indicate the resistance. Instead there are colored bands which are used to specify the resistance. Figure 39-2 explains the color codes and contains two examples.

2. Determine the resistances of R_{AB}, R_{CD}, R_{EF} from the color code markings on each resistor. Determine the tolerance. Record these values in Table 39-1.

3. Determine the true resistance of each resistor by using the appropriate resistance scale on an ohmmeter. Record these values in Table 39-1 and compare them with the values determined in step 2. Indicate whether they are within the tolerance?

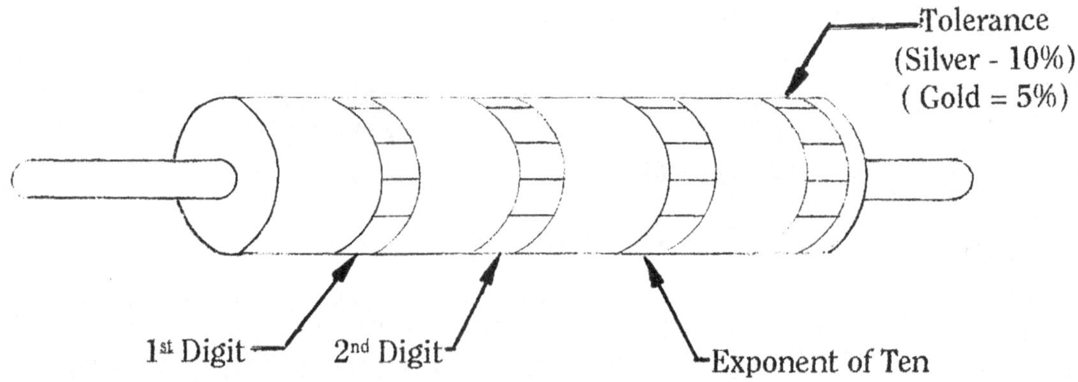

0 black	4 yellow	7 violet
1 brown	5 green	8 gray
2 red	6 blue	9 white
3 orange		

The resistance = (1st digit x 10 + 2nd digit) x 10^{EXP}

Examples

(a) The first, second and third bands are all orange and the last band is gold. The resistance is (30 + 3) x 10^3 ohms = 33,000 ohms with a tolerance of 5% (i.e, The resistance is between 31,350 and 34,650 ohms)

(b) The first band is red, the second band is black, the third band is brown, and the last band is silver. The resistance is 20 x 10^1 ohms = 200 ohms, and the tolerance is 10% (i.e., it falls between 180 and 220 ohms).

Figure 39-2 Resistor Color Codes

4. Assemble the circuits as shown in Figure 39-1. Do not plug the power supply into an electrical outlet until your instructor checks your circuit.

5. Plug in the power supply and turn the dial on the power supply until the milliammeter reads 1 milliampere.

6. Connect the voltmeter across all the resistors, from A to F. Record your reading, V_{AF}, in Table 39-2.

7. Connect the voltmeter from A to B. Record your reading, V_{AB}, in Table 39-2.

8. Connect the voltmeter from C to D. Record your reading, V_{CD}, in Table 39-2.

9. Connect the voltmeter from E to F. Record your reading, V_{EF}, in Table 39-2.

10. Determine the sum of the potential differences $V_{AB} + V_{CD} + V_{EF}$ in Table 39-2.

11. Compare the potential difference across the combination of resistors from step 6 with the sum of the potential differences of the individual resistors from step 10. Consider the value in step 6 as the standard and find the percent error? Record your results in Table 39-2.

12. From Ohm's Law we get I = V/R . Calculate the current through R_{AB} by using Ohm's Law; i.e., $I_{AB} = V_{AB} / R_{AB}$. Use the resistance which was obtained with the ohmmeter from Table 39-1. Record this value in Table 39-3.

13. Calculate $I_{CD} = V_{CD}/ R_{CD}$. Use the resistance which was obtained with the ohmmeter from Table 39-1. Record this value in Table 39-3.

14. Calculate $I_{EF} = V_{EF} / R_{EF}$. Use the resistance which was obtained with the ohmmeter from Table 39-1. Record this value in Table 39-3.

15. Compare I_{AB} , I_{CD}, and I_{EF} with the current through the milliammeter. Calculate the percent error for each and record in Table 39-2. Use the current through the milliammeter as the standard.

16. Calculate R_{eq} from equation (39-1). Use the measured values in Table 39-1 to calculate R_{eq}. Record in Table 39-4.

17. Pull out the electrical plug from your power supply but do not change any settings. Disconnect the resistance board and in its place put the decade box. Turn the dials to a setting which equals R_{eq} determined in step 16. Have your instructor check the circuit before you plug the power supply back into an outlet.

18. Read the current from the milliammeter. Record in Table 39-4 and compare it with the current read from the milliammeter in Table 39-3. Find the percent error. Use the value in Table 39-3 as the standard.

19. What is the voltage across the decade resistance box? Compare it with V_{AF} in Table 39-2 and find the percent error. Use the value from Table 39-2 as the standard.

Questions

1. Why did we use the measured values of the resistances, rather than the values read from the resistor color code, when we calculated the currents through the resistors? (Hint: What does a silver or gold strip mean?)

2. Why were you told not to change any settings when the resistance board was replaced with the decade resistance box?

Name________________________ Date Performed________________________

Lab. Partners ________________________ Date Submitted________________________

____

____

Table 39-1
Determination of the Resistances

	Resistance Form The Color Code	Indicated Tolerance	Measured Resistance	Is the Measured Resistance Within Tolerance? (Yes or No)
R_{AB}	________	________	________	________
R_{CD}	________	________	________	________
R_{EF}	________	________	________	________
$R_{AB} + R_{CD} + R_{EF} =$	________	________	________	________

Table 39-2
Determination of the Potential Differences

$V_{AF} =$ ________ $V_{AB} =$ ________ $V_{CD} =$ ________ $V_{EF} =$ ________

$V_{AB} + V_{CD} + V_{EF} =$ ________________ Percent Error = ____________

$V_{Power\ Supply} =$ ________________ Percent Error = ____________

Table 39-3
Determination of the Currents

	Current (A)	Percent Error
$I_{AB} = V_{AB} / R_{AB}$	____________	____________
$I_{CD} = V_{CD} / R_{CD}$	____________	____________
$I_{EF} = V_{EF} / R_{EF}$	____________	____________
I from milliammeter	____________	____________

Name________________________________

Table 39-4
Determination of the Equivalent Resistance

		Percent Error
$R_{AB} + R_{CD} + R_{EF}$ =	____________________	____________
$V_{Decade\ Box}$ =	____________________	____________
$I_{Decade\ Box}$ =	____________________	____________

EXPERIMENT 40

PARALLEL CIRCUITS

Objective

To study the relationship between the currents and voltages in a parallel circuit.

Equipment

Power supply, milliammeter, voltmeter, ohmmeter, decade resistance box, 3 resistors

Discussion and Theory

We will study the parallel circuit shown in Figure 40-1 and show that:

1. The potential drop across each resistor is equal to the potential difference across the power supply.

2. The current from the power supply equals the sum of the currents through each of the resistors. This is due to *Kirchoff's First Law* which states that *the sum of the currents going into a junction equals the sum of the currents coming out of the junction.* A junction is a point where two or more wires come together.

3. The three parallel resistors R_{AB}, R_{CD}, and R_{EF} can be replaced by an equivalent resistance, R_{eq}, such that

$$\frac{1}{R_{eq}} = \frac{1}{R_{AB}} + \frac{1}{R_{CD}} + \frac{1}{R_{EF}} \qquad (40\text{-}1)$$

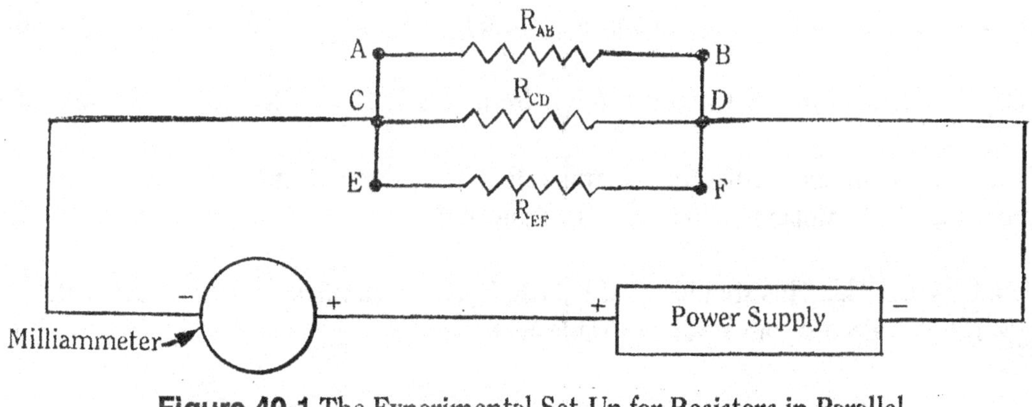

Figure 40-1 The Experimental Set-Up for Resistors in Parallel

Procedure

1. Determine the true resistance of each resistor by using the appropriate resistance scale on an ohmmeter. Record these values in Table 40-1.

2. Assemble the circuit as shown in Figure 40-1. From the positive side of the power supply go to the positive side of the milliammeter. From the negative side of the milliammeter go to terminal C. Connect terminal C to terminal A. Connect terminal C to terminal E. From the negative side of the power supply go to terminal D. Connect terminal D to terminal B and connect terminal D to terminal F. Adjust the power supply so that the milliammeter reads the value that will be given to you by your instructor.

3. Connect the positive side of the voltmeter to terminal A; the negative side to terminal B. Record V_{AB} in Table 40-1.

4. Connect the positive side of the voltmeter to terminal C; the negative side to terminal D. Record V_{CD} in Table 40-1.

5. Connect the positive side of the voltmeter to terminal E; the negative side to terminal F. Record V_{EF} in Table 40-1.

6. Connect the voltmeter across the power supply. The positive side of the voltmeter is connected to the positive side of the power supply; the negative side of the voltmeter is connected to the negative side of the power supply. Record this potential difference V_{PS} in Table 40-1.

7. V_{AB}, V_{CD} and V_{EF} should all be equal to V_{PS}. Use V_{PS} as the standard. Find the percent error in V_{AB}, V_{CD} and V_{Ef} and record in Table 40-1.

8. From Ohm's Law, I = V/R. Calculate the current through the resistor R_{AB} from $I_{AB}=V_{AB}/R_{AB}$. Record I_{AB} in Table 40-1.

9. Calculate the current through the resistor R_{CD} from $I_{CD} = V_{CD} / R_{CD}$. Record I_{CD} in Table 40-1.

10. Calculate the current through the resistor R_{EF} from $I_{EF} = V_{EF} / R_{EF}$. Record I_{EF} in Table 40-1.

11. Read the current in the milliammeter and call it I_{PS}. Convert this result into Amperes. (One milliampere equals 0.001 Amperes). Record I_{PS} in Table 40-1.

12. Calculate $I_{AB} + I_{CD} + I_{EF}$. This should equal I_{PS}. Use I_{PS} as the standard value and calculate the percent error in $I_{AB} + I_{CD} + I_{EF}$. Record your results in Table 40-1

13. Calculate the equivalent resistance, R_{eq} , from equation 40-1. Record this value in Table 40-1.

14. The equivalent resistance of the system is also found from $R_{eq} = V_{PS}/I_{PS}$. Compare this value with the value calculated in step 13 and find the percent error. Use the value found in step 13 as the standard. Record your results in Table 40-1

15. Disconnect the power supply. *Do not alter any of the settings on the power supply.* Remove the resistance board and replace it with a decade resistance box. When wired the circuit should look like Figure 40-2. Set the decade box to the resistance calculated in step 13. Have your instructor check your circuit.

16. Plug in the power supply (with all settings the same as they were before). Read the current from the milliammeter and convert it to Amperes. Record this current in Table 40-1 as I_{PS}'.

15. Find the percent error in I_{PS}'. Use I_{PS} as the standard. Record in Table 40-1.

Questions

1. Three resistors of 5.0 Ohms, 10.0 Ohms and 15.0 Ohms are connected in parallel. What is the equivalent resistance?

2. Three resistors of 30 Ohms each are connected in parallel. What is the equivalent resistance? What is the total current through the circuit if a potential difference of 25 Volts is applied across the three resistors?

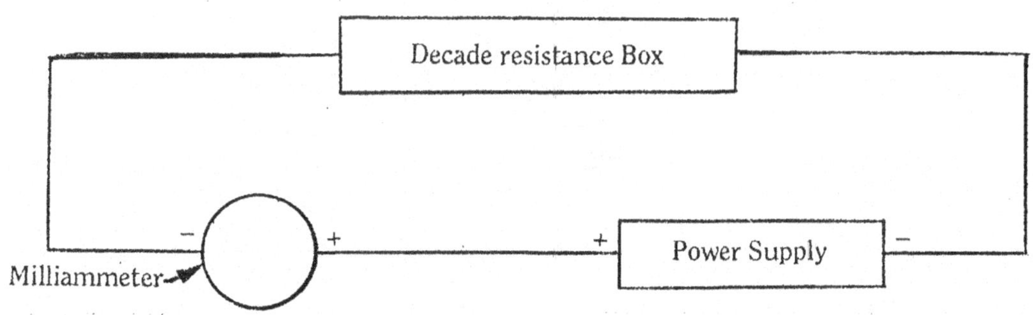

Figure 40-2 Expermental Set-Up With Decade Box Set at R_{eq}

!Name______________________________ Date Performed______________________

Lab. Partners ______________________ Date Submitted______________________

____

____

Table 40-1
Resistors in Parallel

Resistance (ohms)	Voltage (Volts)	Percent Error	Current (Amperes)
R_{AB} = __________	V_{AB} = __________	__________	I_{AB} = __________
R_{CD} = __________	V_{CD} = __________	__________	I_{CD} = __________
R_{EF} = __________	V_{EF} = __________	__________	I_{EF} = __________
	V_{PS} = __________		I_{PS} = __________

$I_{AB} + I_{CD} + I_{EF}$ =______________________________ % Error = ______________________

R_{eq} (From equation (40-1) = ____________

$R_{eq} = V_{PS} / I_{PS}$ =______________________________ % Error = ______________________

I_{PS}' =______________________________ % Error = ______________________

EXPERIMENT 41

THE IDENTIFICATION OF MINERALS

Objective

To identify minerals from their physical properties

Equipment

A set of minerals, steel nail, streak plate (unglazed porcelain tile).

Discussion and Theory

A *mineral* is a naturally occurring homogeneous inorganic crystalline solid that has a fairly consistent chemical composition. All minerals have specific physical properties. These properties are useful for mineral identification. Some of these properties are:

1. *Hardness*: Hardness refers to the ability of one mineral to scratch another. The *Mohs Scale of Hardness* measures the relative hardness of minerals. The scale consists of ten minerals listed in order of increasing hardness. The minerals in the scale are assigned the following values:

Hardness	Mineral
1	Talc (the softest mineral)
2	Gypsum
3	Calcite
4	Fluorite
5	Apatite
6	Orthoclase feldspar
7	Quartz
8	Topaz
9	Corundum
10	Diamond (the hardest mineral)

To use the scale one scrapes a mineral of unknown hardness by each of the Mohs minerals to determine the hardness. Minerals softer then the unknown will not scratch it, but will become scratched themselves. Minerals harder then the unknown will scratch it without being scratched themselves. For example, halite (rock salt) will scratch gypsum (hardness 2) but will not scratch calcite (hardness 3). Therefore, the hardness of halite falls between 2 and 3.

The hardness of some common items that can be used to test for hardness

Hardness	Object
2.5	Fingernail
5	Steel nail
5.5	Glass, steel knife
6	Steel file

2. *Color*: The use of color in identifying minerals could be misleading. In fact, in most instances it is of little value. The presence of small amounts of impurities could cause large variations in coloration. Quartz may be tinted to almost any color. On the other hand, some minerals do have inherent coloration such as sulfur (yellow), azurite (blue) and malachite (green).

3. *Streak*: The color of the mineral when it is powdered is referred to as the streak of the mineral. Streak is determined by scraping (streaking) the mineral across the surface of white unglazed porcelain.

4. *Luster*: How the surface of a mineral reflects light is called its luster. Minerals that reflect light similar to polished metal are said to have a metallic luster. Other common lusters are pearly, vitreous (glassy), silky, earthy (dull) and resinous.

5. *Cleavage*: Cleavage is the tendency of certain minerals to break along certain smooth planes because of planes of weakness in the internal crystal structure. The number of (or lack of) cleavage planes, together with their relative orientation and their degree of perfection, is used for mineral identification.

6. *Fracture*: Minerals that break without exhibiting cleavage are said to fracture. The most common fractures are conchoidal (shell like) fractures which have smooth curved surfaces similar to broken glass, irregular grainy fractures and uneven fractures. Some other types are splintery fractures, even (flat surfaced) fractures and hackly (jagged sharp points) fractures.

7. *Specific Gravity*: The specific gravity is the weight of the mineral divided by the weight of an equal volume of water. Most common rocks have specific gravities around 3 or just a little less. By just handling a mineral it is often possible to determine if it has a relatively high specific gravity (i.e. "it feels heavy"). Galena (a lead ore) has a specific gravity of 7.5. Cinnabar (a mercury ore) has a specific gravity of 8.1.

To determine the specific gravity more accurately the weight of the mineral sample is determined both in air and when fully submerged in water. The specific gravity (S.G.) is then calculated as follows:

$$S.G. = \frac{\text{Weight in air}}{(\text{Weight in air - Weight in water})}$$

$$= \frac{\text{Weight in air}}{\text{Weight loss in water}}$$

8. *Fluorescence and Phosphorescence*: When some minerals are exposed to certain types of electromagnetic radiation (ultraviolet light or x-rays) they emit light. Minerals that emit light while being exposed to radiation are said to be fluorescent. Those minerals that continue to emit light after the exposure to radiation has stopped are said to be phosphorescent.

9. *Other Properties*: Some minerals have unique properties. Some iron-bearing minerals are attracted by a magnet or may be magnetized. The mineral calcite when placed on a book will cause the writing to appear double. Limestone will effervesce when dilute hydrochloric acid is placed on it.

Procedure

For each mineral you are supplied do the following and record your observations in Table 40-2:

1. Indicate its color.

2. Indicate its luster.

3. Indicate its streak.

4. Do not attempt to cleave or fracture any mineral. By observing the mineral indicate if any cleavage planes are present. Indicate the approximate angles between cleavage planes. Indicate if it seems to have fracture.

5. Test for hardness with your fingernail and a nail. Indicate if the hardness is less then 2.5 (soft), between 2.5 and 5 (hard), or greater then 5 (very hard).

6. Indicate any special properties you observe.

7. Compare the physical properties of the minerals studied with the minerals listed in Table 41-1 and identify the minerals you were given.

Table 41-1
Physical Properties of Common Minerals

Mineral Name	Hardness	Specific Gravity	Luster	Common Color	Streak	Cleavage or Fracture	Special Properties
Talc	1	2.75		green to white, gray to white			soapy feel
Graphite	1.5	2.2	Submetallic	black, dark gray		black, dark gray	feels greasy, blackens fingers
Gypsum	2	2.3		white		conchoidal or fibrous	
Galena	2.5	7.4	metallic	gray, black		3 - 90° cleavage planes	
Halite	2.5	2.2	translucent, transparent	white to colorless		3-90° cleavage planes	salty taste
Calcite	3	2.7	vitreous, earthy	white or yellow to colorless		3 non-right cleavages	double refraction effervesces (dilute HCl solution)
Muscovite	2.5	2.9	pearly, vitreous	translucent, colorless to yellow		cleaves into thin sheets	
Biotite	3	2.7	pearly, glassy	dark brown to black		cleaves into thin sheets	
Hematite	2-6.6	5.3	earthy	red, brown	red		
Limonite	2-5.5	3.7	earthy	yellow, black	Yellowish brown		
Augite	5.7	3.5	vitreous	dark green to black	green/gray	2 near 90° cleavages	
Hornblende	5.5	3.3	glassy	dark green to black		56° & 124° cleavages	
Orthoclase	6	2.6	vitreous	white, pinkish-flesh		2 near 90° cleavages	
Plagioclase	6	2.7	vitreous, pearly	white, gray, blue gray		2 near 90° cleavages	striations
Magnetite	6	5.2	metallic	black	black, dark gray		very magnetic
Pyrite	6.3	5		brass yellow, greenish tarnish	greenish brownish black		Cubic crystals common, striations
Quartz	7	2.6	glassy	colorless, white, smoke, rose,violet, any hue		conchoidal fracture	crystals show hexagonal pattern

Name________________________________ Date Performed________________________
Lab Partners__________________________ Date Submitted________________________
____
____

Table 40-2
Identifying Minerals

Mineral Number	Color	Luster	Streak	Fracture or Cleavage Planes	Hardness	Special Properties	Mineral Name